Lecture Notes in Civil Engineering

Volume 242

Series Editors

Marco di Prisco, Politecnico di Milano, Milano, Italy

Sheng-Hong Chen, School of Water Resources and Hydropower Engineering, Wuhan University, Wuhan, China

Ioannis Vayas, Institute of Steel Structures, National Technical University of Athens, Athens, Greece

Sanjay Kumar Shukla, School of Engineering, Edith Cowan University, Joondalup, WA, Australia

Anuj Sharma, Iowa State University, Ames, IA, USA

Nagesh Kumar, Department of Civil Engineering, Indian Institute of Science Bangalore, Bengaluru, Karnataka, India

Chien Ming Wang, School of Civil Engineering, The University of Queensland, Brisbane, QLD, Australia

Lecture Notes in Civil Engineering (LNCE) publishes the latest developments in Civil Engineering - quickly, informally and in top quality. Though original research reported in proceedings and post-proceedings represents the core of LNCE, edited volumes of exceptionally high quality and interest may also be considered for publication. Volumes published in LNCE embrace all aspects and subfields of, as well as new challenges in, Civil Engineering. Topics in the series include:

- Construction and Structural Mechanics
- Building Materials
- Concrete, Steel and Timber Structures
- Geotechnical Engineering
- Earthquake Engineering
- Coastal Engineering
- Ocean and Offshore Engineering; Ships and Floating Structures
- Hydraulics, Hydrology and Water Resources Engineering
- Environmental Engineering and Sustainability
- Structural Health and Monitoring
- Surveying and Geographical Information Systems
- Indoor Environments
- Transportation and Traffic
- Risk Analysis
- Safety and Security

To submit a proposal or request further information, please contact the appropriate Springer Editor:

- Pierpaolo Riva at pierpaolo.riva@springer.com (Europe and Americas);
- Swati Meherishi at swati.meherishi@springer.com (Asia - except China, and Australia, New Zealand);
- Wayne Hu at wayne.hu@springer.com (China).

All books in the series now indexed by Scopus and EI Compendex database!

More information about this series at https://link.springer.com/bookseries/15087

Daniele La Rosa · Riccardo Privitera
Editors

Innovation in Urban and Regional Planning

Proceedings of the 11th INPUT Conference - Volume 2

 Springer

Editors
Daniele La Rosa
DICAR
University of Catania
Catania, Italy

Riccardo Privitera
DICAR
University of Catania
Catania, Italy

ISSN 2366-2557   ISSN 2366-2565   (electronic)
Lecture Notes in Civil Engineering
ISBN 978-3-030-96987-5   ISBN 978-3-030-96985-1   (eBook)
https://doi.org/10.1007/978-3-030-96985-1

© The Editor(s) (if applicable) and The Author(s), under exclusive license
to Springer Nature Switzerland AG 2022
This work is subject to copyright. All rights are solely and exclusively licensed by the Publisher, whether the whole or part of the material is concerned, specifically the rights of translation, reprinting, reuse of illustrations, recitation, broadcasting, reproduction on microfilms or in any other physical way, and transmission or information storage and retrieval, electronic adaptation, computer software, or by similar or dissimilar methodology now known or hereafter developed.
The use of general descriptive names, registered names, trademarks, service marks, etc. in this publication does not imply, even in the absence of a specific statement, that such names are exempt from the relevant protective laws and regulations and therefore free for general use.
The publisher, the authors and the editors are safe to assume that the advice and information in this book are believed to be true and accurate at the date of publication. Neither the publisher nor the authors or the editors give a warranty, expressed or implied, with respect to the material contained herein or for any errors or omissions that may have been made. The publisher remains neutral with regard to jurisdictional claims in published maps and institutional affiliations.

This Springer imprint is published by the registered company Springer Nature Switzerland AG
The registered company address is: Gewerbestrasse 11, 6330 Cham, Switzerland

Preface

A Trans-disciplinary Dialogue for Integrating Nature-Based Solutions in Urban and Regional Planning Processes and Science

The International Conference INPUT2020 on 'Innovation in Urban and Regional Planning'—11th Edition was hosted by the University of Catania (Italy) on 8–10 September 2021 and organised by LaPTA, a research laboratory of the Department of Civil Engineering and Architecture working on sustainable urban and environmental planning.

INPUT is a network of scholars and academics working in different topics of urban and regional planning, particularly focusing on geoinformatics and environmental–ecological aspects of the discipline. Since 1999, INPUT team has been organising and hosting biennial international scientific conferences in Italy, the latest ones taking place in Viterbo (2018), Torino (2016), Napoli (2014), Cagliari (2012) and Potenza (2010), while the next edition of the conference has been scheduled in L'Aquila (Italy) on 2023.

Previously planned on September 2020, due to the COVID-19 pandemic, the INPUT conference was postponed as an hybrid event to 8–10 September 2021, so to promote a more interacting and fruitful meeting among participants. This allowed gathering international scholars in the fields of planning, civil engineering and architecture, ecology and social science, to strengthen the knowledge on nature-based solutions and to enhance the implementation and replication of these solutions in different contexts. About ninety oral presentations by authors from ten countries were grouped into the following fourteen thematic sessions:

- Enhancing the use of nature-based solutions in urban planning
- Modelling to innovate planning solutions for socio-ecological systems
- Input visions: new technologies, data and hybrid models for spatial planning
- Urban metabolism and simulation for decision-making in spatial planning
- Performance-based planning
- Computational planning
- Geodesign for informed collaborative spatial decision-making

- Planning and design of ecosystems services: assessment frameworks, models, mapping and implications
- Green infrastructure for planning healthy urban environments
- The mitigation of peripheralisation risk in urban and regional planning
- Strategies and actions for climate change adaptation and mitigation in Mediterranean regions
- Analysis and planning of rural landscapes
- Accessibility in urban planning: moving towards innovative approaches
- Maintenance, upgrading and innovation in cultural heritage.

The conference focused on how to integrate nature-based solutions in urban and regional planning science and practice. Nature-based solutions have been defined by the International Union for the Conservation of Nature as actions to deploy, manage or restore urban ecosystems to address societal challenges, such as climate change, food security, risk to natural and anthropogenic disasters or social inequalities. NBS provides sustainable, cost-effective, multipurpose and flexible alternatives than more traditional solutions used which are usually based on built infrastructure systems. INPUT 2020 Conference stressed the basic idea that using components that mimic natural processes in the built environment could generate a wide number of benefits in cities and produced more equal, safe and livable urban environment. Papers presented in the conference discussed widely on possible ways to include NBS in planning processes and tools, so to consolidate the so far fragmented evidence that NBS can significantly improve the quality of urban environments.

Following a first call for papers launched in Winter 2020, sixty-nine contributions among the ones submitted to the INPUT 2020 Conference were collected in the book 'Innovation in Urban and Regional Planning – Proceedings of the 11th INPUT Conference – Volume 1 (La Rosa and Privitera, 2021), published on late Summer 2021 in the Springer series 'Lecture Notes in Civil Engineering'. Contributions covered the topics of nature and ecosystems for urban systems, models and technologies for spatial planning, climate change and spatial planning, peripheries, rural and cultural landscapes and accessibility.

On February 2021, a second call for papers was launched, and this allowed to enlarge the audience of the contributors. This book presents a collection of further selected articles, structured in the three following topical parts:

- Nature and ecosystems for urban systems
- Modelling for spatial planning
- Peripheries, rural and cultural landscapes.

The second volume of the INPUT 2020 Conference Proceedings provides additional reflections and proposals on empirical frameworks for NBS. Attitude of NBS to be integrated in urban planning practices for drawing healthier and more livable urban environments has been widely investigated through a leading ecosystem services approach. Computational tools, technologies, data and hybrid models are explored for providing innovative spatial planning modelling methodologies. Furthermore, prospective roles of NBS in planning science and practice are investigated in the light of peripheralisation risks, rural landscapes and innovation in cultural heritage.

December 2021   Daniele La Rosa
                                                                                   Riccardo Privitera

Reference

La Rosa D., Privitera R. (eds) Innovation in Urban and Regional Planning. INPUT 2021. Lecture Notes in Civil Engineering, vol 146. Springer, Cham. https://doi.org/10.1007/978-3-030-68824-0_1

Organisation

Organising Committee

Daniele La Rosa	Laboratorio per la Pianificazione Territoriale e Ambientale, Department of Civil Engineering and Architecture—University of Catania, Italy
Riccardo Privitera	Laboratorio per la Pianificazione Territoriale e Ambientale, Department of Civil Engineering and Architecture—University of Catania, Italy
Luca Barbarossa	Laboratorio per la Pianificazione Territoriale e Ambientale, Department of Civil Engineering and Architecture—University of Catania, Italy
Viviana Pappalardo	Laboratorio per la Pianificazione Territoriale e Ambientale, Department of Civil Engineering and Architecture—University of Catania, Italy
Francesco Martinico	Laboratorio per la Pianificazione Territoriale e Ambientale, Department of Civil Engineering and Architecture—University of Catania, Italy
Paolo La Greca	Laboratorio per la Pianificazione Territoriale e Ambientale, Department of Civil Engineering and Architecture—University of Catania, Italy

Scientific Committee

Ginevra Balletto	University of Cagliari, Italy
Luca Barbarossa	University of Catania, Italy
Ivan Blecic	University of Cagliari, Italy
Dino Borri	Polytechnic University of Bari, Italy
Domenico Camarda	Polytechnic University of Bari, Italy
Michele Campagna	University of Cagliari, Italy
Valerio Cutini	University of Pisa, Italy
Andrea De Montis	University of Sassari, Italy
Romano Fistola	University of Sannio, Italy
Chiara Garau	University of Cagliari, Italy
Carmela Gargiulo	University of Napoli 'Federico II', Italy
Davide Geneletti	University of Trento, Italy
Roberto Gerundo	University of Salerno, Italy
Federica Gobattoni	Tuscia University, Italy
Paolo La Greca	University of Catania, Italy
Daniele La Rosa	University of Catania, Italy
Sabrina Lai	University of Cagliari, Italy
Giuseppe Las Casas	University of Basilicata, Italy
Antonio Leone	University of Salento, Italy
Giampiero Lombardini	University of Genova, Italy
Beniamino Murgante	University of Basilicata, Italy
Raffaele Pelorosso	Tuscia University, Italy
Alessandro Plaisant	University of Sassari, Italy
Riccardo Privitera	University of Catania, Italy
Bernardino Romano	University of L'Aquila, Italy
Francesco Scorza	University of Basilicata, Italy
Maurizio Tira	University of Brescia, Italy
Angioletta Voghera	Polytechnic University of Turin, Italy
Corrado Zoppi	University of Cagliari, Italy
Francesco Zullo	University of L'Aquila, Italy

Contents

Modelling for Spatial Planning

Peripheries, Rural and Cultural Landscapes

About the Editors

Daniele La Rosa (PhD in Urban and Regional Planning) is Associate Professor of Urban and Environmental Planning at the Department Civil Engineering and Architecture of the University of Catania (Italy). He teaches spatial planning and urban design in Building Engineering MSc course at the University of Catania.

His research interests include sustainable urban planning, ecosystem services, GIS applications for urban and landscape planning, environmental indicators, environmental strategic assessment, land use science and landscape studies. He is Author of more than 90 publications on the above-mentioned topics and Member of editorial board of relevant international peer-review journals, such as Nature-Based Solutions (Elsevier) and Socio-Ecological Practice Research (Springer).

Riccardo Privitera is Assistant Professor in Urban and Spatial Planning at the Department of Civil Engineering and Architecture, University of Catania (Italy). He holds a PhD degree in Urban Planning, taught Urban Design as Lecturer in Architecture MSc programme and got the Italian National Scientific Qualification as Associate Professor in Urban and Regional Planning.

He is Member of the Italian Centre of Urban planning Studies, Member of the European Land-use Institute, Visiting Academic Researcher at the Department of Urban Studies and Planning University of Sheffield (UK) and Visiting Professor at the Faculty of Science University of Alexandria (Egypt). He has been working on several urban local plans for municipalities, provincial administrations and universities, such as land use plans, regeneration of historical centres plans and landscape protection plans. As Member of the coordination staff, he has been working on many European and UK projects.

Among his scientific interests in the field of urban and regional planning, sustainable urban growth, non-urbanised areas planning, urban green infrastructures, urban quality, green cities, climate change adaptation and mitigation strategies, ecosystem services, land cover analysis, urban morphology analysis, land suitability analysis, urban and peri-urban agriculture, farmland protection, real estate development processes, transfer of development rights, renewable energy sources and energy efficiency issues at urban scale are included. Based on these topics, he published more than fifty among scientific papers, proceedings, chapters and books.

Nature and Ecosystems for Urban Systems

Assessing Potential for and Benefits of Scaling up Nature-Based Solutions in Malmö

Chiara Cortinovis[✉], Peter Olsson, Niklas Boke-Olén, and Katarina Hedlund

Centre for Environmental and Climate Science, Lund University, Lund, Sweden
chiara.cortinovis@cec.lu.se

Abstract. While many projects have demonstrated the potential of single Nature-based Solutions (NbS) to contribute to urban climate change adaptation, the challenge now lies in moving from demonstration projects to a full-scale deployment. The aim of this research is to assess the potential for a full-scale implementation of NbS in Malmö (Sweden), and the expected benefits and co-benefits. We developed six scenarios that simulate the current condition, the full-scale implementation of different NbS strategies (i.e., installing extensive green roofs, planting street trees, desealing parking areas, and enhancing vegetation in urban parks), and a combination of them. Then, we assessed the scenarios in terms of heat mitigation, stormwater regulation, carbon storage, biodiversity potential, and overall greenness, using a combination of spatially-explicit methods. Overall, the research reveals that the impacts of scaling up NbS depends on two factors: i) the existing potential to integrate NbS in the urban fabric, and ii) the capacity of each NbS type to deliver benefits in certain conditions.

1 Introduction

In the last few years, the concept of Nature-based Solutions (NbS) has become increasingly popular to designate actions that take advantage of nature to address societal challenges in a sustainable way (Lafortezza and Sanesi 2019). Several policy initiatives, especially in the EU, are promoting NbS as cost-effective strategies to sustainably address a variety of urban challenges (Faivre et al. 2017), including climate change adaptation (Kabisch et al. 2017).

Many exemplary projects have demonstrated the potential role of single, small- and medium-scale NbS to adapt cities to climate change (Sapundzhieva et al. 2020). NbS such as green roofs and rain gardens increase water retention and infiltration, thus helping to prevent urban flooding due to increasingly intense rain events (Haghighatafshar et al. 2018). Street trees and urban forests contribute to reduce air temperature through shading and evapotranspiration, thus limiting the negative impacts of more frequent and intense heatwaves (Zardo et al. 2017).

The challenge now lies in moving from demonstration projects to a full-scale deployment of NbS. This requires "making the case" for NbS by acting on regulatory frameworks, business and financial models, and the societal acceptance of these solutions (Frantzeskaki et al. 2019). However, intrinsic factors also limit the potential to scale up

© The Author(s), under exclusive license to Springer Nature Switzerland AG 2022
D. La Rosa and R. Privitera (Eds.): INPUT 2021, LNCE 242, pp. 3–11, 2022.
https://doi.org/10.1007/978-3-030-96985-1_1

NbS in cities, a critical one being the availability of space, especially in existing high-density built-up areas (Grace et al. 2021). To set realistic targets for scaling up NbS, policy-makers need to know the amount of different NbS types that can fit in the urban space and the impacts that can be expected from their implementation.

This research aims to assess the potential for a full-scale implementation of NbS for climate change adaptation in Malmö (Sweden), and the expected benefits and co-benefits. To this aim, we developed six NbS implementation scenarios, including the current condition, and, for each scenario, we assessed two climate change-related benefits-heat mitigation and stormwater regulation–and three additional co-benefits.

The city of Malmö, home of around 350,000 people, is the third largest city in Sweden and the main urban center in the southern part of the country, also due to its strategic connection to Copenhagen. The vulnerability to climatic events such as intense rainfalls and heat waves (Sörensen and Emilsson 2019; Rocklöv and Forsberg 2009), combined with the orientation towards densification strategies as a response to population growth (Malmö's residents are expected to rise to 500,000 within 30 years) (Malmö Stad 2018), make it an interesting case to assess the potential of NbS for climate change adaptation of existing urban areas.

2   Materials and Methods

2.1   Developing NbS Scenarios

We developed six spatially-explicit scenarios that simulate the effects of different climate change adaptation strategies based on NbS. One scenario represents the current condition, four scenarios focus on scaling up a single NbS, and a last one simulates the combined implementation of the four strategies (Table 1).

Operationally, NbS implementation is simulated through land cover transitions, i.e. we applied GIS-based algorithms that translate a set of rules about NbS size and location into land cover changes. First, we mapped the current land cover into eight classes that include different vegetation typologies as well as green roofs. We used a combination of remote-sensing data (ortophotos and derived products) and existing datasets (coastline and building footprints), and obtained land cover maps at a resolution of 1 m. Then, to model land cover transitions based on homogeneous spatial information, we included additional data from the Urban Atlas and from OpenStreetMap.

The scenarios were designed by taking into account the constraints to scaling up NbS determined by space availability and technical feasibility. Additional economic, social, and institutional aspects potentially affecting the process of NbS implementation were not considered. To simulate ambitious but feasible interventions, we did not include any action that involves a change of existing land uses or the removal of buildings and transportation networks.

2.2   Assessing Benefits and Co-benefits of NbS Implementation

The two climate change adaptation benefits were quantified through the InVest v 3.8.7 'Urban Cooling' and 'Urban Flood Risk Mitigation' models (Sharp et al. 2020). The

Table 1. NBS implementation scenarios simulated in the study (* based on OpenStreetMap data, ** based on OpenStreetMap and Urban Atlas data).

Scenario	Strategy	Land cover transition rules
Current	–	Current land cover
GreenRoof	Installing green roofs	Extensive green roofs are installed on all roofs with size above 40 m^2 and slope below 20° (calculated based on NDSM)
ParkingAreas	De-sealing parking areas	Existing parking areas* are de-sealed and converted into concrete-reinforced lawns; trees existing on the areas are maintained
Parks	Enhancing vegetation in urban parks	Within existing parks**, part of the areas currently sealed (excluding paths*, sport fields*, allotments*, and cemeteries*) is converted into low vegetation; the tree coverage is increased by adding a tree every 100 m^2 of plantable area
StreetTrees	Planting street trees	Trees are planted along secondary streets* and residential roads*, whenever enough space is available (we assumed 4 and 2 lanes respectively)
GreenDream	All of the above	All of the above

potential of NbS to lower high (summer) temperatures in the city was assessed using the heat mitigation index, which accounts for four factors: albedo, crop coefficient (evapotranspiration), canopy coverage (shading), and the cooling effect generated by large green areas on their surroundings. We used land cover maps to calculate canopy coverage, while crop coefficient and albedo for the current scenario were retrieved from the Landsat-based EEFLUX (https://eeflux-level1.appspot.com/), selecting Aug 12, 2020 as a typical summer day (low cloud cover, warm temperature, low wind, and low water availability). In the other scenarios, the values corresponding to land cover changes were adjusted by using the median of the respective new land cover class over the city. The model was run using the standard parameters, considering each block as a separate land use class.

The potential of NBS to reduce stormwater runoff was assessed through the runoff retention index, which measures the share of stormwater that is retained in the analysed area. The InVest model applies the Curve Number (CN) method developed by the USDA. We assigned CN values to the different land cover classes based on the standard values provided by USDA (NRCS 1986), assuming all impermeable surfaces connected to the drainage system. Maps of hydrological soil groups with a resolution of 250 m were generated based on the maps of saturated hydraulic conductivity in the 3D Soil Hydraulic

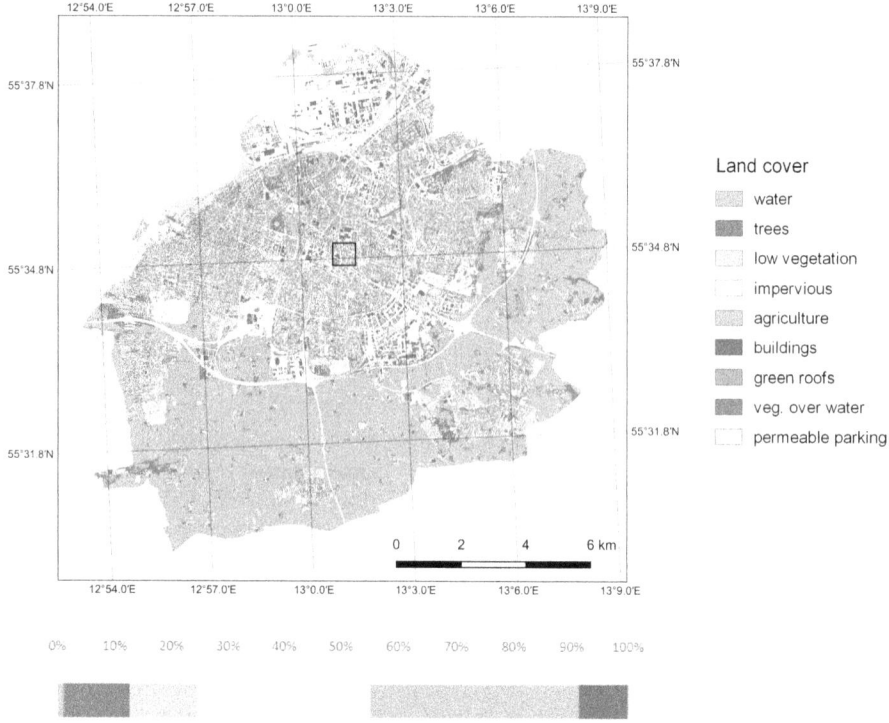

Fig. 1. Map and share of land covers in Malmö for the *Current* scenario. The black rectangle indicates the enlargement shown in Fig. 2.

Database of Europe (Tóth et al. 2017) combined with the map of depth to bedrock in the soilGrids 250 m (Shangguan et al. 2017). We run the model considering a rain event of 20 mm, which is a frequent rain in Malmö and one for which the relative effects of NbS implementation are the greatest (compared to more intense cloudburst).

Besides the two main benefits, we selected three additional co-benefits that cover aspects related different socio-environmental urban challenges: carbon storage, linked to climate change mitigation; biodiversity potential, linked to biodiversity conservation; and overall greenness, linked to the health benefits of a green environment for the resident population. Carbon storage was modeled as a function of land cover, adopting values retrieved from the literature. Biodiversity potential at the block level was calculated through an index based on structural diversity and share of green area (Radford and James 2013). Overall greenness was measured as the share of green and blue areas in a 500-m neighborhood of each point.

3 Results

3.1 Potential for Scaling up NbS in Malmö

Malmö is considered, not erroneously, a green city, with more than 60% of the area covered by "green" land cover classes (including water, trees, low vegetation, and agriculture) (Fig. 1). The urban structure is predominantly concentric, with a dense urban core surrounded by diffused low-density residential neighborhoods, extensive commercial and industrial areas, and agricultural fields (36% of the area). The presence of large commercial and industrial areas, and of the harbor to the North, determines the high rate of soil sealing: each 1 m^2 of building footprint corresponds to more than 3.5 m^2 of impervious surfaces. Many urban parks, including some large ones, are distributed across the city and cover a total area of more than 1,450 ha. On average, they are not intensely planted (37.5% tree cover), which reflects the preference for lawns and open areas in urban green spaces.

These starting conditions affect the potential for NbS implementation simulated in the scenarios (Fig. 2). The area involved in land cover changes ranges from 1% to 10% of the city area in the *ParkingAreas* and *GreenDream* scenarios, respectively (Table 2). Green roofs are the single NbS with the highest potential in terms of area involved (more than 5% of the city area, 61% of existing building footprints). Enhancing vegetation in existing urban parks – as simulated in the *Parks* scenario-leads to land cover changes involving around 2.5% of the city area, with a significant increase in tree cover (+16%) and decrease in the share of impervious surfaces (−4%) over the city. The *StreetTrees* scenario results in the addition of more than 52,000 new trees, leading to an increase in tree coverage of more than 12% compared to the current condition (Table 2).

Table 2. Changes simulated in the NbS implementation scenarios compared to the *Current* scenario.

Scenario	Change area (ha)	Main changes
GreenRoof	816 (5.2%)	61% building area covered by green roofs
ParkingAreas	159 (1.0%)	−3.2% impervious
Parks	399 (2.5%)	+16.1% tree cover, −4.0% impervious
StreetTrees	224 (1.4%)	+12.1% tree cover, −4.5% impervious
GreenDream	1,588 (10.0%)	+27.9% tree cover, −11.5% impervious, 61% building area covered by green roofs

3.2 Benefits of NbS Implementation Scenarios

As expected, the *GreenDream* scenario, which combines the four NbS types, is the best performing across all benefit indicators (Fig. 3). After the *GreenDream*, the ranking of the scenarios depends on the analyzed indicator. Considering the two benefits related to climate change adaptation, the *Parks* scenario ranks second in heat mitigation, while the

Fig. 2. Enlargement of the land cover map of Malmö under the different scenarios.: 1) *Current*, 2) *GreenRoofs*, 3) *ParkingAreas*, 4) *Parks*, 5) *StreeTrees*, 6) *GreenDream*. For the legend, please refer to Fig. 1.

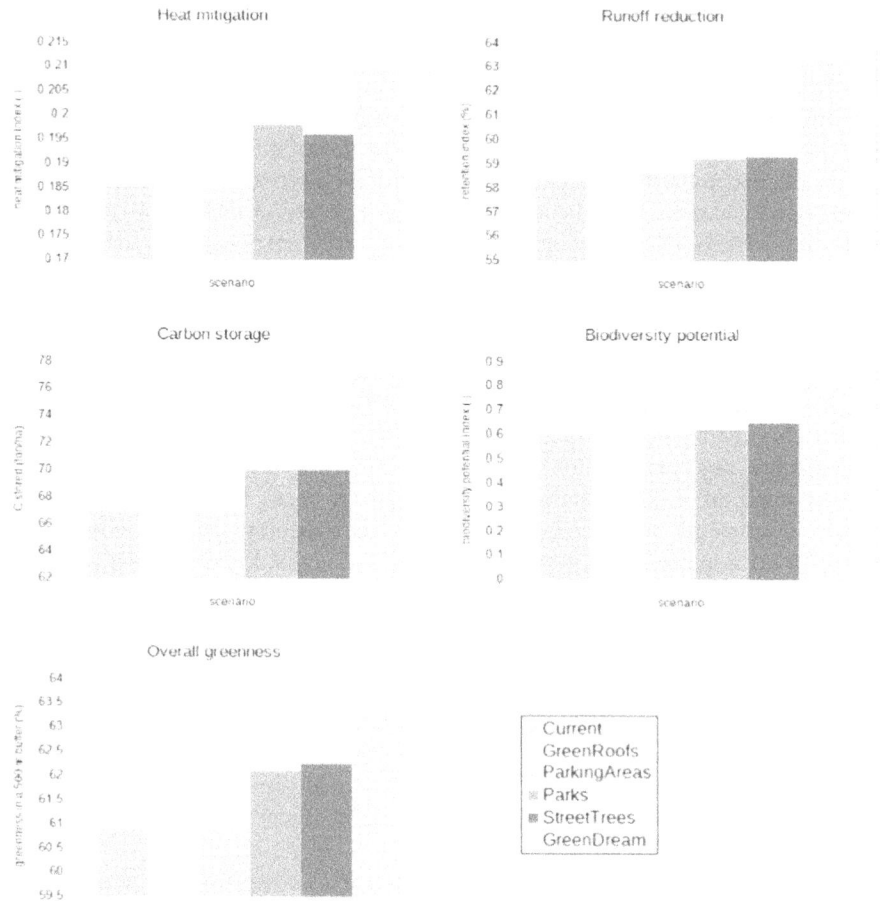

Fig. 3. Comparison of benefits and co-benefits under the different NbS implementation scenarios (average values across the city).

GreenRoofs scenario is the second best in stormwater regulation. The latter also produces the greatest increase in biodiversity potential, but has no or little effect on heat mitigation and overall greenness. While the *GreenRoofs*, *Parks,* and *StreetTrees* scenarios produce comparable effects on carbon storage, the *StreetTrees* scenario outclasses the other two in terms of overall greenness.

4   Discussion and Conclusions

The results reveal that most NbS provide multiple benefits. This evidence is coherent with the literature on NbS in urban contexts (Kabisch et al. 2016) and justifies policies supporting NbS implementation to exploit synergies in the provision of urban ecosystem services (European Commission 2015). Furthermore, the results at the city scale do not highlight any trade-off among the analyzed benefits and co-benefits. Negative effects on biodiversity potential and greenness were observed only in a few cases, at the level of single blocks. This is coherent with a view of NbS as win-win strategies and, more in general, with the scientific literature, which shows limited trade-offs among urban ecosystem services (Howe et al. 2014). Rather, trade-offs in cities usually emerge with other land uses. In Malmö there is already a strong pressure for development, which is expected to rise in the next years. The selection of NbS types accounted for this, as no undeveloped land is occupied, hence the scenarios are future-proof from this perspective.

Different NbS implementation scenarios show a different potential to address different challenges. Among the strategies applying a single NbS type, installing green roofs is the best one to reduce runoff and increase biodiversity, while planting more trees -either along streets or in existing urban parks– produces the greatest impact on heat mitigation and greenness. However, when comparing these results, the amount of changes simulated in each scenario should also be considered. For example, the area converted into green roofs is twice the area of new tree cover added in existing parks. Moreover, we did not include any consideration about the financial feasibility of NbS implementation. Different types of intervention have different unitary costs, can be financed through different instruments, and correspond to different actors bearing the costs of implementation. All these factors are key to determine the real possibility of achieving the level of implementation simulated in the scenarios.

Aggregated values at the city scale provide only a partial view of the impacts of NbS implementation. An example is the heat mitigation index, which shows a very limited increase in the average value across the city, but a much more significant local increase in some areas directly affected by the simulated interventions. Looking at the spatial distribution of the benefits and co-benefits, and their matching with the distribution of population and vulnerability factors (Cortinovis and Geneletti 2019), is therefore critical to assess the impacts of scaling up NbS and to direct pubic investments where NbS benefits are more needed.

Overall, the research reveals that the benefits and co-benefits of scaling up NbS depend on two factors: the existing potential to integrate NbS in the urban fabric, and the capacity of each NbS type to deliver benefits in certain conditions. Both factors should be considered when defining goals and strategies for NbS implementation.

Acknowledgments. This research received funding from the Horizon 2020 project NATURVA-TION (grant agreement 730243) and from the FORMAS project 'Nature-based solutions for urban challenges' (n. 2016-00324).

References

Cortinovis, C., Geneletti, D.: A framework to explore the effects of urban planning decisions on regulating ecosystem services in cities. Ecosyst. Serv. **38**, 100946 (2019)

Faivre, N., Fritz, M., Freitas, T., de Boissezon, B., Vandewoestijne, S.: Nature-Based Solutions in the EU: innovating with nature to address social, economic and environmental challenges. Environ. Res. **159**, 509–518 (2017)

Frantzeskaki, N., et al.: Nature-based solutions for urban climate change adaptation: linking science, policy, and practice communities for evidence-based decision-making. Bioscience **69**(6), 455–466 (2019)

Grace, M., et al.: Priority knowledge needs for implementing nature-based solutions in the Mediterranean islands. Environ. Sci. Policy **116**, 56–68 (2021)

Haghighatafshar, S., Nordlöf, B., Roldin, M., Gustafsson, L.G., la Cour, J.J., Jönsson, K.: Efficiency of blue-green stormwater retrofits for flood mitigation–conclusions drawn from a case study in Malmö, Sweden. J. Environ. Manage. **207**, 60–69 (2018)

Howe, C., Suich, H., Vira, B., Mace, G.M.: Creating win-wins from trade-offs? Ecosystem services for human well-being: a meta-analysis of ecosystem service trade-offs and synergies in the real world. Glob. Environ. Change **28**, 263–275 (2014)

Kabisch, N., et al.: Nature-based solutions to climate change mitigation and adaptation in urban areas - perspectives on indicators, knowledge gaps, barriers and opportunities for action. Ecol. Soc. **21**(2), 39 (2016)

Kabisch, N., Korn, H., Stadler, J., Bonn, A.: Nature-Based Solutions to Climate Change Adaptation in Urban Areas: Linkages Between Science, Policy and Practice. Springer, Cham (2017). https://doi.org/10.1007/978-3-319-56091-5

Lafortezza, R., Sanesi, G.: Nature-based solutions: settling the issue of sustainable urbanization. Environ. Res. **172**, 394–398 (2019)

Malmö Stad: Comprehensive Plan for Malmö (2018). https://malmo.se/download/18.6c44cd5c1 7283283332b3de/1592233669232/OP_english_summary_lores.pdf. Accessed 1 June 2021

NRCS: Urban Hydrology for Small Watersheds. TR-55 (1986)

Radford, K.G., James, P.: Changes in the value of ecosystem services along a rural–urban gradient: a case study of Greater Manchester, UK. Landscape Urban Plan. **109**, 117–127 (2013)

Rocklöv, J., Forsberg, B.: Comparing approaches for studying the effects of climate extremes - a case study of hospital admissions in Sweden during an extremely warm summer. Glob. Health Action **2**(1), 034 (2009)

Sapundzhieva, A., et al.: ReNature: creating the first nature-based solutions compendium in the Mediterranean. Res. Ideas Outcomes **6**, e59646 (2020)

Shangguan, W., Hengl, T., Mendes De Jesus, J., Yuan, H., Dai, Y.: Mapping the global depth to bedrock for land surface modeling. J. Adv. Model Earth Syst. **9**, 65–88 (2017)

Sharp, R., et al.: InVEST 3.8.7 User's Guide (2020)

Sörensen, J., Emilsson, T.: Evaluating flood risk reduction by urban blue-green infrastructure using insurance data. J. Water Res. Pl **145**(21), 04018099 (2019)

Tóth, B., Weynants, M., Pásztor, L., Hengl, T.: 3D soil hydraulic database of Europe at 250 m resolution. Hydrol. Process. **31**, 2662–2666 (2017)

Zardo, L., Geneletti, D., Pérez-Soba, M., Van Eupen, M.: Estimating the cooling capacity of green infrastructures to support urban planning. Ecosyst. Serv. **26**, 225–235 (2017)

Integrating Nature-Based Solutions into Urban Planning and Policies: Learning from the Apulia Case Study

Angela Barbanente and Laura Grassini[✉]

Polytechnic University of Bari, Via Edoardo Orabona, 4, 70126 Bari, Italy
{angela.barbanente,laura.grassini}@poliba.it

Abstract. Much research has been done to evaluate the impacts of Nature-based Solutions (NBS) in spatial planning and management at the different scales and to show their positive effects. Main policy documents also mention the importance to adopt nature-based approaches in urban planning, built on circular economies, integration and multifunctionality. Nevertheless, when it comes to actual practices in urban planning and policies, approaches and solutions still mainly follow old traditional approaches. How to foster the uptake of NBS innovations in practice is thus a crucial question for urban planners.

The paper aims at contributing to this discussion by investigating whether, to what extent and in what form NBS have succeeded in piercing in the Apulia region, where some recent policies have been trying to foster the uptake of such solutions first into urban regeneration projects and then into the broader urban planning sector. To this purpose the framework known as Multi-Level Perspective (MLP) has been applied to show the transition pathway undertaken in urban planning and policies in Apulia towards an NBS approach as well as the potential of different policy instrument mixes to support and give long-term perspectives to nature-based innovations while lowering resistance to their consolidation and diffusion.

Keywords: Nature based solutions · Urban planning · Apulia · Transitions · Multi-level perspective

1   Key Challenges for NBS in Urban Contexts

NBS are solutions inspired and supported by nature, which are locally adapted, resource-efficient and systemic; they simultaneously provide environmental, social and economic benefits and help build resilience [1]. The EU has promoted them as innovations to solve multiple societal challenges – like climate change, disaster risk reduction, water and food security, human health [2] – and to contribute to the fulfilment of the UN sustainable development goals [3]. They have thus been identified as a core pillar in the EU Research and Innovation Programme and in several EU policies like the European Green Deal, the EU biodiversity strategy for 2030 and the new climate adaptation strategy, the Green infrastructure strategy and the Urban Agenda for the EU. They constitute a cornerstone in

© The Author(s), under exclusive license to Springer Nature Switzerland AG 2022
D. La Rosa and R. Privitera (Eds.): INPUT 2021, LNCE 242, pp. 12–21, 2022.
https://doi.org/10.1007/978-3-030-96985-1_2

major international policy agreements, too, like the Paris Agreement on climate change, the new urban agenda[1], the forthcoming Post-2020 Global Biodiversity Framework.

Challenges they contribute to face are particularly striking in cities, which are considered at the same time the places where the impacts of unsustainable development produce the most severe effects, with the highest costs borne by the most vulnerable, and the places which have the highest potential to boost change in the transitions towards sustainable futures [4]. Because of this, NBS and Green Infrastructures are identified, by the Urban Agenda for the EU, as a core instrument to tackle main priorities and crosscutting issues for action in urban areas. In particular, they are mentioned in relation to the "climate adaptation" priority – where green infrastructure solutions may contribute to anticipating the adverse effects of climate change and to prevent or minimise the damage caused to urban areas – and to the "sustainable use of land" priority – where NBS may ensure sustainable land use changes in both growing and shrinking cities as well as in renaturing/greening of urban areas.

According to the report of the Expert Group on NBS and Re-Naturing Cities [1], their main contributions in urban areas are expected to be towards: i) enhancing sustainable urbanisation, thus making cities more attractive, increasing well-being and stimulating economic growth; ii) restoring degraded ecosystems, thus improving the resilience of ecosystems, enabling them to deliver vital ecosystem services and to meet other societal challenges; iii) developing climate change adaptation and mitigation; iv) improving risk management and resilience in more effective ways thanks to synergies in reducing multiple risks. Based on those contributions, several key R&I areas have been identified for further actions, which include urban regeneration, well-being in urban areas, coastal resilience, multi-functional watershed management and ecosystem restoration, use of matter and energy, insurance value of ecosystems, carbon sequestration.

The scientific evidence base related to the effectiveness of NBS in urban areas is rapidly expanding in Europe [5, 6], despite the persistent uncertainty on the degree of causality and effectiveness due to overlapping of multiple stressors and complex synergies between different domains [7]. Large attention is thus currently put to enlarge the focus from single solutions to scale-up [8] and to the development and application of comprehensive assessment frameworks for NBS in urban areas as well as to the analysis of their potential trade-offs [5, 9, 10].

Instead, still limited research seems to address other problematic issues and knowledge gaps like governance and institutional arrangements for NBS implementation in urban areas [11, 12], where a multiplicity of stakeholders need to be involved in decision-making processes [13] and particular attention needs to be paid to empowering citizens and local communities in co-production and co-management of solutions [14, 15].

Another under-investigated issue relates to mainstreaming the NBS approach into urban planning and policies, which are still largely informed by traditional development approaches and incorporate NBS as "technological fixes". Limited analysis has been carried out on broader system transitions and on the diffusion of innovation beyond innovation niches [16, 17].

In the attempt to deal with these knowledge gaps, the paper analyses the transition pathway undertaken in urban planning and policies in the Apulia region, where some

---

[1] In this case, the term "nature-based innovation" has been used.

recent policies have been trying to foster the uptake of NBS first into urban regeneration initiatives and then into the broader urban planning sector. For the analysis, the Multi-level Perspective (MLP) framework is applied. Building on the lessons learned from the undergoing transition, some concluding remarks are made on how to support the mainstreaming of the NBS approach in urban planning. These reflections may be relevant in view of the coming large investments planned by the EU in NBS in connection with the Green Deal and the post-pandemic Recovery and Resilience Plans.

2   Framework for the Analysis of Transitions in Urban Planning and Policies Towards NBS Approaches

Over the past two to three decades a new body of research has emerged focusing on sustainability transitions in so-called socio-technical systems (STS), i.e. those systems encompassing co-evolutionary dynamics of change among technological, socio-cultural, economic, institutional and policy dimensions [18]. Several approaches have been developed to deal with them [19], ranging from strategic niche management (STM), to Technological Innovation Systems (TIS) to Multi-Level Perspective (MLP).

These frameworks have been widely applied to several fields. Yet, only few studies have started applying them to transitions connected to NBS with the exception of applications of TIS frameworks [17]. NBS mainstreaming into urban planning and policy has been analysed only recently through the application of frameworks derived from studies of other environmental policies [20], or through the use of change agents' models [16]. Nevertheless, these frameworks are unable to explain the complex causality and the multi-level and multi-sectoral dynamics of change.

Developing a line of research aimed at investigating the dynamics of innovation and the conditions for its acceleration and stabilisation in urban policies and planning [21, 22], we found it worthwhile to apply the MLP framework to the study of transitions in urban planning towards NBS approaches. MLP has been developed in the broad field of innovation studies based on insights from evolutionary economics, from sociology of technology and from neo-institutional theory. It frames transitions in socio-technical systems as the result of co-evolutionary and non-linear dynamics of change taking place within and across three levels. The lowest level ("niche innovations") provides "incubation rooms" for radical novelties [23]. The intermediary level of "socio-technical regime" then embodies the deep-structural rules, cognitive routines and beliefs that coordinate and guide perceptions and actions of actors belonging to different system dimensions, i.e. science, technology, culture, industry policy and market/user preference. Finally, the upper level of "socio-technical landscape" represents the wider exogenous context of macro-economic trends, deep cultural patterns, macro-political development, etc., which influences niche and regime dynamics while being beyond the control of individual actors.

According to this framework, socio-technical transitions develop through the alignment of processes at different levels. When niche experimentations become aligned and stabilized within a dominant design, if appropriate "windows of opportunity" are opened up at the regime level thanks to pressures from the landscape level, innovations may break through [24] and create changes in the socio-technical regime.

Some MLP scholars are recently investigating the conditions under which socio-technical transitions can be deliberately accelerated [25, 26] and the influence policy instrument mixes can exert on the rate and direction of transitions [27]. These works have started shedding light on the way deliberate strategies may work, on one side, to encourage the breakthrough of socio-technical systems from initial niches and, on the other side, to lower the resistance to change from incumbent actors and powerful regime players [27].

The application of the MLP framework to the analysis of the transition pathway undertaken in the Apulia region for NBS mainstreaming in urban planning involves analysing changes along different dimensions: a technical dimension made of different types of planning instruments at different scales (strategic vs. regulatory plans, regeneration vs. expansion plans, guidelines for sustainable general and detailed plans ...), user practices and cultural meanings (how aware local communities are of climate change and disaster risk, how much they are involved in planning interventions, ...), institutional structures (which governance frameworks can foster the development and implementation of NBS, how they interact with existing government structures, ...), policy (what legislative, regulatory, strategic, financial and organisational measures are developed to direct territorial transformations), markets (how inclined to innovation are private firms active in urban transformation, ...), scientific and technical knowledge (which skills and competencies professionals involved in plan making have, and how they can develop new technical competencies required by innovative NBS, what role scientific research institutions can play in this respect, ...), and infrastructures (how new green and blue infrastructures interact with grey infrastructural networks at the different urban scales, whether and to what extent they are considered additional or alternative to these, ...).

3 Learning from the Apulia Case Study

In the Apulia region, the "socio-technical regime" is strongly anchored to a spatial planning system essentially based on (mostly old) municipal master plans that include large areas of agricultural land to be developed for housing, industry or commercial activities, and at most provide for the protection of cultural heritage, natural areas and landscapes of exceptional value. Replacing the established planning system with a new one is far from easy. Moreover, this is likely to be ineffective when such a system, as it is in our case study, combines rigid and detailed regulations with a tendency to disregard the rules. Aware of this, since 2006 the regional government has been promoting "innovation niches" aimed at directing such system towards the ecological, social and economic regeneration of the built environment through a mix of financial, regulatory, organisational and relational instruments[2]. The underlying assumption was that fostering NBS as part of a public policy aimed to promote biodiversity and make urban areas more climate resilient could not be based only on laws and other formal acts, but required a

---

[2] This classification is based on the well-known NATO taxonomy of policy instruments referred to the government resources used, i.e.: nodality (advice, training, information, ...), authority (legal power), treasury (grants, incentives, public funding,...), organization (government human resources, ability to action change) [28].

mix of instruments involving the different actors that contribute to put these solutions into practice.

Figure 1 provides a picture of the transition dynamics developing in the case study, according to the framework illustrated in Sect. 2. Along the y-axis the three levels of niche innovations, socio-technical regime and landscape are represented, while the x-axis represents time. The clouds of small arrows moving from niche experimentation upwards and the dashed arrows indicating the landscape pressures exerted by the different levels of government—EU, national and regional—show the multilevel and multisectoral dynamics of change that seek to break the socio-technical regime by acting on the six dimensions of market, institutions, industry, science, policy, culture, and technology.

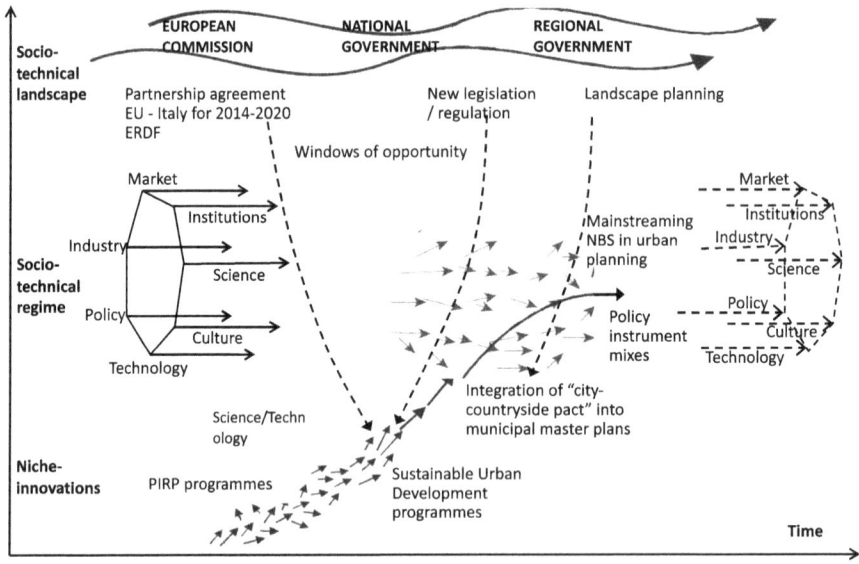

Fig. 1. Transition pathway in the Apulia region for NBS mainstreaming in urban planning (own elaboration on [24]).

The first innovation niche was the "Integrated Programme for the rehabilitation of peripheral neighbourhoods" (PIRP), which funded 122 municipal initiatives for the regeneration of deprived and degraded areas [26] using both regional funds and the ERDF Regional Operational Programme (ROP) 2007–2013. Its call for proposals emphasised the sustainable use of resources and NBS, providing rewards for programmes addressing environmental sustainability objectives and natural resources protection, reduction of water consumption and improvement of soil permeability, reduction of energy consumption and use of renewable energy sources, improvement of the thermal inertia of buildings, etc. In order to mainstream the PIRP approach to urban regeneration, the regional law no. 21/2008 introduced the Integrated urban regeneration programme, a new instrument to be used for promoting urban regeneration through a coordinated set of interventions that, among others, include "the rehabilitation of the urban environment through the provision of ecological infrastructure such as green and blue networks designed to increase biodiversity in the urban environment... spaces with a high degree of permeability...", as well as "the ecological regeneration of settlements aimed at saving resources, with particular reference to soil, water and energy, ... increasing the provision of ecological infrastructure and encouraging sustainable mobility".

Up to this stage, policy instruments supported the implementation of NBS, although they did not explicitly mention NBS and their potential to address climate change adaptation and disaster risk reduction. Then, the regional regulation n. 2753/2010 required the adoption of NBS in the design of detailed urban plans, based on the application of bioclimatic criteria. On the other hand, in the 2014–2020 ERDF-ESF ROP the objectives of "Climate change adaptation and risk prevention and management" were explicitly included in the "Priority Axis XII—Sustainable Urban Development"[3]. Thus, the call for proposal for urban rehabilitation projects funded by this Axis specified that 20% of the resources assigned to each project would be allocated to actions for biodiversity protection and enhancement and green infrastructures as a contribution to adaptation to climate change, and explicitly encouraged proponents to include interventions for permeabilizing and greening artificial surfaces, improving the urban microclimate, and developing green infrastructures.

This change, aimed at boosting NBS mainstreaming in urban planning and policies, was supported not only by lessons learned from the implementation of regeneration programmes funded under the previous programming cycle, where a "soft" inclusion of NBS within the call objectives had led to limited implementation[4], but also by the "window of opportunity" opened up thanks to pressure exerted by the landscape level: the Thematic Objective (TO) "Climate change adaptation and risk prevention and management" was included among the 11 TOs of the partnership agreement between Italy and the EU for the 2014–2020 ERDF programming cycle.

Further impetus for the mainstreaming of NBS in local plans and projects came from the regional Territorial Landscape Plan (TLP), started in 2008 and approved in 2015. In this plan, NBS take on a different meaning: they are part of a wide-ranging strategy that aims at re-establishing the co-evolutionary relationships between human settlement and nature for a self-sustainable local development [25]. In this perspective, the focus of urban regeneration is not limited to target neighbourhoods, but involves the widest relationships between city and countryside, built environment and open spaces.

The territorial project "city-countryside pact" and its guidelines are a pillar of this strategy. They aim to promote an agro-urban and agro-environmental policy for the landscape, in an attempt to bind the urban and the rural in a new alliance. The boundary space between city and countryside is the place where the potential for NBS is clearer. The pact provides for the preservation and rehabilitation of these agricultural areas still present within and on the edges of the urban in a multifunctional way: enhancing their ecological

---

[3] The focal point of this Axis is on strengthening sustainability in its social and ecological dimensions.

[4] This is because key players of the existing socio-technical regime had a low propensity to innovate, particularly with regard to the ecological regeneration of the target areas.

and hydrogeological functions, biodiversity protection and water conservation, as well as their socio-cultural, recreational and therapeutic functions.

The integration of the strategic vision of the "city-countryside pact" into municipal master plans is thus triggering new approaches to urban planning at large, with the point of view of multifunctionality offering a reversal of perspective with respect to usual planning practices. In particular, public space can be multifunctional in that it performs more than one function at the same time, related to hydrogeological safety, water and energy saving, education and recreation.

The implementation of "city-countryside pact" is still ongoing through the municipal master plans' compliance process with the TLP, which may allow in due course the stabilization of the NBS approach in urban planning[5]. In the meanwhile, aware of the long-time needed for this process of formal adaptation of old municipal plans, the regional government has adopted a serial of policy instrument mixes to support this transition across all dimensions of the system. One of this is the use of part of the 2014–2020 ERDF-ESF's Axis VI on "Environmental protection and promotion of natural and cultural resources" to support the implementation the "city-country pact", through local projects aimed to improve the ecological quality and social use of transition spaces between the urban landscape and the open countryside. This further boost for innovation proved particularly effective in breaking the regime by requiring professional skills and competencies different from those involved in established urban planning and design practices. Cycle of seminars and online forums for professionals and public officials to disseminate methodologies and best practices, as well as the promotion of processes of social production of the territory and development of proactive engagement of local communities, further contributed to the purpose.

4   Conclusions

In this paper we have analysed the ongoing transition pathway undertaken in urban planning and policies in Apulia towards an NBS approach. The regional level of analysis has been considered to be relevant for this purpose due not only to the important intermediary role the regional government can play in facilitating or hindering effective transitions at town and city level, but also to the multiple government resources it can activate to foster them.

In Apulia, after the promotion of several innovation niches in the field of urban regeneration, "windows of opportunity" have been opened at the regime level due to pressures from the landscape level reflecting principles and resolutions, developed in technical and policy documents from the Commission and other European agencies, to support NBS for climate change adaptation and disaster risk reduction. This led to the definition of specific objectives and conditionalities supporting NBS in the allocation of cohesion policy funds to urban regeneration projects in the period 2014–2020.

But such episodes cannot produce by themselves a quantum change into ordinary urban planning practices, unless a set of purposive policy instrument mixes developed

---

[5] This possible outcome is suggested by some of the about 30 urban plans where the compliance has been already completed or is at an advanced stage – e.g. San Severo, Campi Salentina, Fragagnano, Melpignano.

by the regional level is recognized to foster transition across several dimensions. In the Apulia case, the ordinary statutory tools (like Regional Law 21/2008, Regulation n. 2753/2010, TLP) proved to be essential, as they acted to support innovation niches and to contrast established practices in the regime.

However, transition is still ongoing. The lesson learned from the Apulia experience is that integrating NBS into urban planning and policies requires a long time to be fully assimilated in local practices, as it involves deep changes in scientific, technological, socio-cultural, economic, institutional and policy dimensions, which need to be supported by adequate instrument mixes. In order to ease the integration of NBS into planning processes a crucial role can be played by regional legislation, regulations and landscape planning insofar as they constitute an integrated system of guidelines and incentives to guide urban transformations towards NBS, and are supported by multidisciplinary training programs for professionals in the public and private sectors and the promotion of NBS culture among authorities and population.

The coming large investments planned by the EU for the Green Deal and the post-pandemic Recovery and Resilience Plans will open a large window of opportunities for the implementation of NBS, as 38% of the Next Generation EU funds allocated to Italy will be devoted to "green transition" projects. But as transition studies warns us, this is not sufficient to guarantee replacement of the old regime, especially if the emphasis of EU evaluation criteria will be on efficiency of spending and the longer time required to draft innovative NBS will be labelled as "inefficiency". The combination of instrument mixes would thus be essential to support the effective breakthrough potential of forthcoming opening of windows of opportunities.

References

1. European Commission: Towards an EU Research and innovation agenda for nature-based solutions and renaturing cities. CEC, Brussels (2015)
2. Cohen-Shacham, E., Walters, G., Janzen, C., Maginnis, S.: Nature-based solutions to address global societal challenges. International Union for Conservation of Nature, Gland, Switzerland (2016). https://portals.iucn.org/library/sites/library/files/documents/2016-036.pdf
3. Faivre, N., Fritz, M., Freitas, T., de Boissezon, B., Vandewoestijne, S.: Nature-Based Solutions in the EU: innovating with nature to address social, economic and environmental challenges. Environ. Res. **159**, 509–518 (2017)
4. European Commission: Urban Agenda for EU 'Pact of Amsterdam' (2016). https://ec.eur opa.eu/regional_policy/sources/policy/themes/urban-development/agenda/pact-of-amster dam.pdf
5. Raymond, C.M., Berry, P., Breil, M., Nita, M.R., Kabisch, N., et al.: An impact evaluation framework to support planning and evaluation of nature-based solutions projects. Report prepared by the EKLIPSE Expert Working Group on Nature-based Solutions to Promote Climate Resilience in Urban Areas. Centre for Ecology & Hydrology, Wallingford (UK) (2017)
6. European Environment Agency: Urban adaptation in Europe: how cities and towns respond to climate change. EEA Report No. 12/2020. Publications Office of the EU, Luxembourg
7. Kabisch, N., Frantzeskaki, N., Pauleit, S., Naumann, S., Davis, M., et al.: Nature-based solutions to climate change mitigation and adaptation in urban areas: perspectives on indicators, knowledge gaps, barriers, and opportunities for action. Ecol. Soc. **21**(2), 39 (2016). https://doi.org/10.5751/ES-08373-210239

8. Frantzeskaki, N., Borgstrom, S., Gorissen, L., Egermann, M., Ehnert, F.: Nature-based solutions accelerating urban sustainability transitions in cities. In: Kabisch, N., Korn, H., Stadler, J., Bonn, A. (eds.) Nature-Based Solutions to Climate Change Adaptation in Urban Areas – Linkages between Science, Policy and Practice. Springer, Cham (2017). https://doi.org/10.1007/978-3-319-56091-5

9. Dooling, S.: Ecological gentrification: a research agenda exploring justice in the city. Int. J. Urban Reg. Res. **33**, 621–639 (2009). https://doi.org/10.1111/j.1468-2427.2009.00860.x

10. Wolch, J.R., Byrne, J., Newell, J.P.: Urban green space, public health, and environmental justice: the challenge of making cities "just green enough". Landsc. Urban Plan. **125**, 234–244 (2014)

11. Frantzeskaki, N.: Seven lessons for planning nature-based solutions in cities. Environ. Sci. Policy. **93**, 101–111 (2019)

12. Mendes, R., Fidélis, T., Roebeling, P., Teles, F.: The institutionalization of nature-based solutions—a discourse analysis of emergent literature. Resources **9**, 6 (2020). https://doi.org/10.3390/resources9010006

13. van Ham, C., Klimmek, H.: Partnerships for nature-based solutions in urban areas – showcasing successful examples. In: Kabisch, N., Korn, H., Stadler, J., Bonn, A. (eds.) Nature-based Solutions to Climate Change Adaptation in Urban Areas. TPUST, pp. 275–289. Springer, Cham (2017). https://doi.org/10.1007/978-3-319-56091-5_16

14. Crowe, P.R., Foley, K., Collier, M.J.: Operationalizing urban resilience through a framework for adaptive co-management and design: five experiments in urban planning practice and policy. Environ. Sci. Policy. **62**, 112–119 (2016)

15. Frantzeskaki, N., Kabisch, N.: Setting a knowledge co-production operating space for urban environmental governance: lessons from Rotterdam, Netherlands and Berlin. Germany. Environ. Sci. Policy. **62**, 1–9 (2016)

16. Kronenberg, J., Bergier, T., Maliszewska, K.: The challenge of innovation diffusion: nature-based solutions in Poland. In: Kabisch, N., Korn, H., Stadler, J., Bonn, A. (eds.) Nature-based Solutions to Climate Change Adaptation in Urban Areas. TPUST, pp. 291–305. Springer, Cham (2017). https://doi.org/10.1007/978-3-319-56091-5_17

17. van der Jagt, A.P.N., Raven, R., Dorst, H., Runhaar, H.: Nature-based innovation systems. Environ. Innov. Soc. Transit. **35**, 202–216 (2020)

18. Geels, F.W., Schot, J.W.: The dynamics of transitions: a socio-technical perspective. In: Grin, J., Rotmans, J., Schot, J.W. (eds.) Transitions to Sustainable Development: New Directions in the Study of Long-Term Transformative Change. Routledge, London (2010)

19. Markard, J., Raven, R., Truffer, B.: Sustainability transitions: an emerging field of research and its prospects. Res. Policy **41**, 955–967 (2012)

20. Wamsler, C., Pauleit, S., Zölch, T., Schetke, S., Mascarenhas, A.: Mainstreaming nature-based solutions for climate change adaptation in urban governance and planning. In: Kabisch, N., Korn, H., Stadler, J., Bonn, A. (eds.) Nature-based Solutions to Climate Change Adaptation in Urban Areas. TPUST, pp. 257–273. Springer, Cham (2017). https://doi.org/10.1007/978-3-319-56091-5_15

21. Barbanente, A., Grassini, L.: Transitions towards landscape- and heritage-centred local development strategies: a multi-level perspective. In: AESOP 2019 Conference—Book of Papers 'Planning for Transition', Venice, 9–13 July 2019

22. Barbanente, A., Grassini, L.: Fostering innovation in area-based initiatives for deprived neighbourhoods: a multi-level approach. Int. Plan. Stud. **25**(2), 206–221 (2020)

23. Schot, J.W.: The usefulness of evolutionary models for explaining innovation. The case of the Netherlands in the nineteenth century. History Technol. **14**, 173–200 (1998)

24. Geels, F.W.: Technological transitions as evolutionary reconfiguration processes: a multi level perspective and a case study. Res. Policy **31**, 1257–1274 (2002)

25. Roberts, C., Geels, F.W., Lockwood, M., Newell, P., Schmitz, H., et al.: The politics of accelerating low-carbon transitions: towards a new research agenda. Energy Res. Soc. Sci. **44**, 304–311 (2018)
26. Roberts, C., Geels, F.W.: Conditions and intervention strategies for the deliberate acceleration of socio-technical transitions: lessons from a comparative multi-level analysis of two historical case studies in Dutch and Danish heating. Technol. Anal. Strateg. Manag. **31**(9), 1081–1103 (2019)
27. Kivimaa, P., Kern, F.: Creative destruction or mere niche support? Innovation policy mixes for sustainability transitions. Res. Policy. **45**, 205–217 (2016)
28. Hood, C.: The Tools of Government. Chatham House Publishers, Chatham (1986)

Integrating Blue-Green Infrastructure into Dense Urban Watersheds in Istanbul for Increased Flood Resilience

Pinar Pamukcu-Albers[1]([✉]), Betul Uygur Erdogan[2], Dilek Eren Akyuz[3], Hasret Sahin[4], and Mehmet Ali Derse[5]

[1] Nature Conservation Centre, Ankara, Turkey
pinarpamukcu@hotmail.com
[2] Department of Watershed Management, Faculty of Forestry, Istanbul University-Cerrahpasa, Istanbul, Turkey
uygurb@iuc.edu.tr
[3] Department of Civil Engineering, Faculty of Engineering, Istanbul University-Cerrahpasa, Istanbul, Turkey
dilekeren.akyuz@iuc.edu.tr
[4] Department of Energy Systems Engineering, Faculty of Technology, Gazi University, Ankara, Turkey
hasret.sahin@gazi.edu.tr
[5] Department of Landscape Architecture, Faculty of Architecture, Cukurova University, Adana, Turkey
maderse@cu.edu.tr

Abstract. The irreversible change in the land cover due to anthropogenic activities has accelerated the climate change impacts on cities. This acceleration has also created unpredictable stress such as flood-related hazards because of the increment in impervious surfaces. Moreover, the sudden and unpredictable changes in the precipitation pattern complicate sustainable water management. This situation directs the authorities to find nature-based and cost-effective solutions. Hence, this study offers a novel approach for flood risk assessment using climate change projections. The objectives of this study are to: (1) determine flood risk based on multi-criteria decision analysis in three dense urban watersheds in Istanbul, (2) explain the change in the future extreme events using climate projections, and (3) recommend blue-green infrastructure to decision-makers for the prioritized high flood risk areas. The global climate models MPI-ESM-LR, IPSL-CM5A-MR, and MOHC-HadGEM2-ES and their regional models were used considering RCPs 4.5 and 8.5 scenarios. The time ranges were set as of 2011–2015 (past), 2051–2055 (near-future) and 2091–2095 (future). The results showed that in the future period, based on MPI-ESM-LR global climate model projections for RCP 4.5 scenario, total precipitation increment will be a maximum of 1.70 times higher according to the past period.

Keywords: Blue-green infrastructure · Flood risk · Climate projections

© The Author(s), under exclusive license to Springer Nature Switzerland AG 2022
D. La Rosa and R. Privitera (Eds.): INPUT 2021, LNCE 242, pp. 22–28, 2022.
https://doi.org/10.1007/978-3-030-96985-1_3

1 Introduction

Blue-green infrastructure (BGI) concept has become popular for decision-makers because they help to improve the components of cities providing ecosystem services (de Oliveira Rolo et al. 2021). Therefore, their planning and design have been performed to make cities more resilient to the impacts of both hazards and climate change (Kapetas and Fenner 2020).

Within this context, this study presents a novel approach for flood risk assessment involving BGI suggestions for Istanbul, which is considered as a vulnerable city to climate change impacts (Aygün and Baycan 2018). Previous studies typically only investigated the hydrological risks and water management strategies in Istanbul (Kömüşcü and Çelik 2013; van Leeuwen and Sjerps 2016). However, according to the authors' knowledge, the "blue/green needs & solutions" nexus considering the future climate projections have not been studied so far for Istanbul. Considering the gap in the literature, this study has the following objectives: (1) determine flood risk based on Analytic Hierarchy Process, which is a multi-criteria decision-making model in three watersheds in Istanbul, (2) explain the change in the future extreme events using climate projections, and (3) recommend blue-green infrastructure to decision makers for the prioritized high flood risk areas.

2 Materials and Methods

2.1 Study Area and Data Acquisition

The study area (about 582 km^2) is composed of three watersheds (from west to east; Sazlidere, Alibeykoy and Kagithane, respectively), which are between 41° 28'–41° 03' N and 29° 59'–29° 02' E on the European side of Istanbul Province (Fig. 1). Impervious areas drive up every single day due to urbanization in the watersheds where part of Istanbul's water comes from surface water and where it is collected. Therefore, the impacts of urbanization on natural processes and ecosystems such as natural hazards can be exemplified in this area.

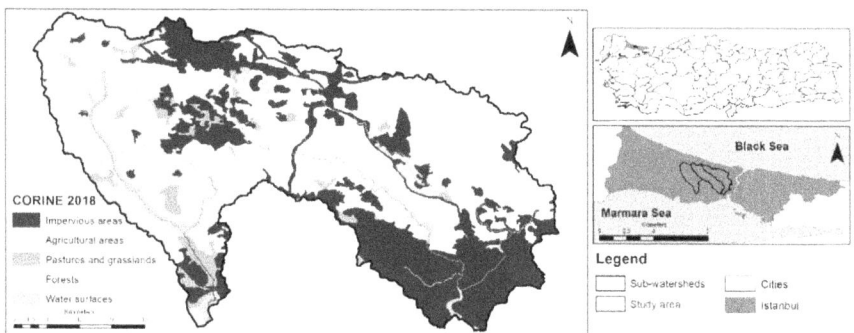

Fig. 1. Study area, Istanbul, Turkey.

The digital elevation model (DEM) provided by the Advanced Spaceborne Thermal Emission and Reflection Radiometer was used to determine borders of watersheds and streams, and to analyze spatial DEM-derived factors (slope, aspect, and distance to rivers). The soil map was obtained from the Ministry of Agriculture and Forestry of the Republic of Turkey. The annual average total precipitation (1938–2010) of three meteorology stations (Terkos, Florya, and Sariyer) in Istanbul were utilized to determine the distribution of precipitation in the watersheds. In assessing the flood risk, land use and land cover were included in the analysis with the reclassification of the CORINE Land Cover 2018. The important buildings were acquired from the Open Street Map. Moreover, three global climate models' (GCMs) processed data (MPI-ESM-LR, IPSL-CM5A-MR, and MOHC-HadGEM2-ES) for RCPs 4.5 and 8.5 scenarios in the periods of 2011–2015, 2051–2055 and 2091–2095 were used for climate projections.

2.2   Flood Risk Assessment

Analytic Hierarchy Process (AHP), a multi-criteria decision-making method, was preferred to assess dissemination of the flood risk and to determine flood-prone areas. The pairwise comparison matrix was built on using seven parameters (slope, soil, aspect, distance to rivers, precipitation, land use and flow). All factors and sub-factors were integrated into the GIS environment using ArcGIS 10.1. The defined sub-factors were classified in raster format, and then the raster calculator was used to overlay weighted factors. The sub-factors were weighted based on literature. The evaluation scale from 1 to 5 was used for the weights of sub-factors. In this case, the sub-factor with the highest risk had 5 and the one with the lowest risk had 1. The weights of main factors were delivered using a questionnaire (Goepel 2018; URL1). The questionnaire was opened online on well-recognized web-based and social media platforms. One hundred experts, specialized in watershed management and urban planning, participated in the survey. The standard AHP linear scale method was applied because of the low consistency ratio (Pagano et al. 2021).

A high slope accelerates the flow in any land use/land cover. There is more flood risk for the downstream of the basin due to its flatness (Kazakis et al. 2015), so when a low slope means a high risk of exposure to floods. The slope factor was classified by 5 groups (<2%, 2–5%, 5–10%, 10–15%, and >%15) considering flood susceptibility in percent (Vasconcelos et al. 2017). The soil factor was classified according to Hydrologic Soil Groups considering the drainage and slope groups of national soil groups (Öztürk 2009). The aspect factor affects the amount of runoff. In this study, aspect was distinguished into 4 categories (Flat areas, North-Northeast-Northwest, South-Southeast-Southwest, and East-West). Another factor used was the distance to the rivers and was classified by 5 groups (0–200 m, 200–500 m, 500–1000 m, 1000–2000, and >2000 m) according to the study by Raja et al. (2017). Furthermore, the national strict preservation zone of rivers (200 m) is also considered. It has been assumed that the flood risk will decrease as moving away from streams. The dissemination of precipitation was carried out using a type of deterministic method for multivariate interpolation technique "Inverse Distance Weighting" and Thiessen polygons were used to analyze the proximity and neighborhood of the stations. The precipitation factor was classified using natural breaks in 5 groups (794.19–826.80 mm, 762.21–794.19 mm, 731.48–762.21 mm,

703.88–731.48 mm, and 667.51–703.88 mm). Land use was classified by 4 groups using the Level 3 of CORINE land cover: Forests; Impervious areas; Pastures and grasslands; and Agricultural areas. The amount of runoff was calculated by the SCS-CN method (Woodward et al. 2003) for forests with good and fair hydrological conditions pastures, agricultural areas, impervious areas and urban green areas, and classified by 5 groups using natural breaks (84.60–100%, 62.84–84.60%, 45.44–62.84%, 28.37–45.44%, and 14.64–28.37%).

The prioritization was given to the flood-prone areas according to the intended use of the buildings (museums, hospitals, schools, governmental institutions, universities, and stadiums) and parks. Thus, it is aimed to make these important buildings and parks usable during and after the flood periods. Priority buildings and parks susceptible to flooding were determined by overlaying the layers of flood-prone areas and the simulations of the daily mean precipitation flux for the near-future and future periods. In the context of BGI, flood-prone areas have been highlighted not only for the priority buildings/areas, but also for the areas causing the floods.

2.3 Climate Scenarios

CORDEX regional climate model data on single levels from Copernicus Climate Database was selected to perform the analysis. The simulations of the daily mean precipitation flux were assessed for two scenarios (RCPs 4.5 and 8.5) for past (2011–2015), near-future (2051–2055), and future (2091–2095) periods using three global climate models (MPI-ESM-LR, IPSL-CM5A-MR and MOHC-HadGEM2-ES), together with two regional models (MPI-CSC-REMO2009 and SMHI-RCA4) to determine the change in precipitation amount. The mean values for each month were calculated, and then the near-future and future predictions of precipitation flux ($kgm^{-2}s^{-1}$) were divided by the models' data of precipitation in the past period.

3 Results and Discussion

3.1 Flood-Prone Areas

According to the results of the questionnaire, the weights of factors were determined as 21% for precipitation, 20.2% for slope, 14% for soil, 13% for distance to rivers, 12.1% for land use, 11.3% for runoff, and 8.4% for aspect. The reclassify tool was used to provide a new range of values to normalize the output values to be between 1 and 100. Flood susceptibility divided by 5 groups based on these values: 1–20 (Very low), 21–40 (Low), 41–60 (Moderate), 61–80 (High), and 81–100 (Very high). The distribution of flood risk in the study area is given in Fig. 2.

Flood risk assessment results showed that 39.61% (223.40 km^2) of the total study area has moderate flood risk. Although these moderate flood risk areas spread over the study area, they are aggregated in a large part of the Sazlidere watershed and in the south of the Kagithane watershed. Areas with moderate flood risk are followed by the areas with low (157.36 km^2), high (83.77 km^2), very low (83.56 km^2) and very high (15.97 km^2) flood risks, respectively. High and very high flood risk areas are 17.68% of

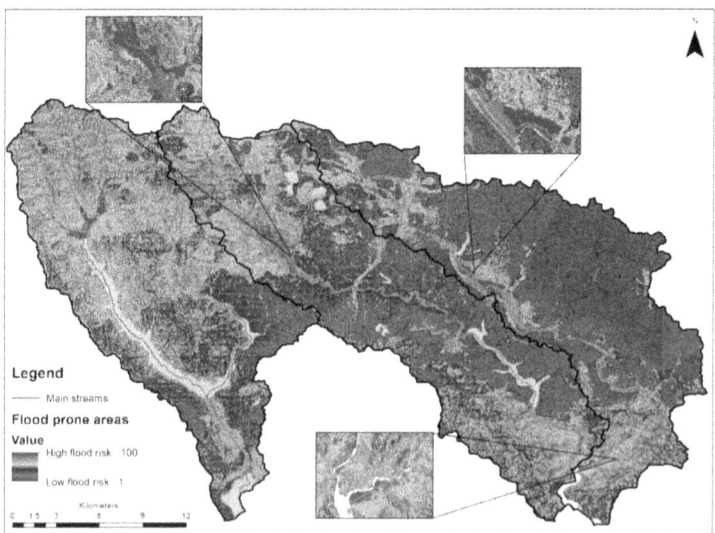

Fig. 2. Flood-prone areas in the study area.

the study area. These areas are especially floodplains and have impervious areas with increasing the flood susceptibility in the study area. 188 existing important buildings (inc. museums, hospitals, schools, governmental institutions, universities, and stadiums) and 186 parks are located in the moderate flood risk areas. On the other hand, 38 buildings and 47 parks are within the area classified as high flood risk. 4 buildings and 5 parks, included in the very high flood risk area, are located on the banks of the streams.

3.2 Comparison of Past, Near-Future and Future Climate Scenarios

The month of December was chosen for the analysis since it has the highest precipitation amount in Istanbul (MGM 2021). According to model projections, precipitation will drive up in the near-future (2051–2055) in the range of 0.64–1.18 times for RCP 4.5 scenario and in the range of 0.52–1.07 times for RCP 8.5 scenario based on the total precipitation rate in December for the past period (2011–2015). The increment of precipitation in the future period (2091–2095) will be in the range of 0.69–1.70 based on RCP 4.5 scenario and in the range of 0.56–1.05 based on RCP 8.5 scenario. The analysis showed that MPI-ESM-LR global climate model projection has the highest increment in precipitation in the future period (Fig. 3). Total precipitation increment will be a maximum of 1.70 times higher for RCP 4.5 scenario according to the past period. For RCP 8.5 scenario, the increment in the total precipitation was in the range of 0.56–0.79. For the near-future period based on RCPs 4.5 and 8.5 scenarios, increment in the total precipitation remains in the range of 0.61–0.93.

As shown in Fig. 3, there are universities, schools, hospitals, town halls and many parks in areas with high and very high flood risk, especially in the south of the Kagithane watershed, where the urban density is the highest in the study area. According to the MPI-ESM-LR GCM projection based on RCP 4.5 scenario for future period in this part

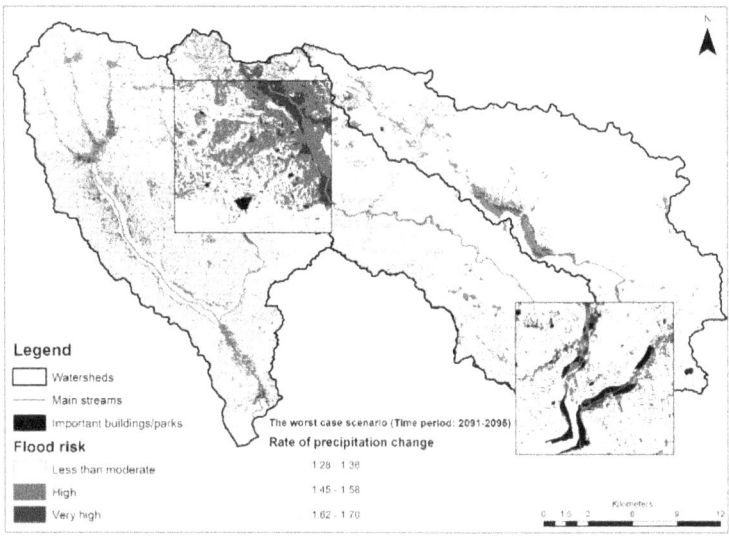

Fig. 3. Flood-prone areas together with the worst-case scenario in the future period (2091–2095) for December.

of the watershed, the increment of total precipitation will be in the range of 1.28–1.58 times higher compared to the past period. Upstream of the Alibeykoy watershed, where discontinuous urban fabric is located, schools and parks are in the flood-prone areas. For this part of Alibeykoy watershed, the increment in the total precipitation in the future period was in the range of 1.55–1.70 based on the MPI-ESM-LR GCM projection for RCP 4.5 scenario.

4   Conclusion

The main finding of the study is that the flood risks of three watersheds in Istanbul (namely, Sazlidere, Alibeykoy and Kagithane) will be increased by 1.28–1.70 times based on MPI-ESM-LR GCM projection for RCP 4.5, which is the worst scenario. Considering this dramatic increment, it is still questionable whether the existing water management infrastructure in the flood-prone areas will be sufficient to handle this expected excessive precipitation in the future. At this point, blue-green infrastructure solutions should carry into effect especially for public spaces such as hospitals, schools, and parks. The impervious surfaces of public spaces should be replaced with pervious materials and turned into integrated green spaces where rainwater storage can be implemented. Moreover, establishing large-scale rain gardens will be very effective to reduce the speed and the amount of surface runoff in areas having high slopes and impervious surfaces. These types of BGI solutions not only help to promote water efficiency on a small scale but also improve urban environmental health and sustainability in Istanbul. To preserve biodiversity and sustainability in the climate change adaptation process, the climate-resilient of Istanbul should be strengthened considering its high population and accelerated urbanization. The local authorities in Istanbul should implement a holistic approach while developing solutions for the flood-prone areas and develop future-oriented strategy spatial plans in micro-scale levels.

References

Aygün, A., Baycan, T.: Istanbul's vulnerability to climate change: an urban sectors' based assessment. In: Leal Filho, W., Manolas, E., Azul, A.M., Azeiteiro, U.M., McGhie, H. (eds.) Handbook of Climate Change Communication: Vol. 3. CCM, pp. 361–383. Springer, Cham (2018). https://doi.org/10.1007/978-3-319-70479-1_23

de Oliveira Rolo, D., Gallardo, A.L.C.F., Ribeiro, A.P., Siqueira-Gay, J.: Local society perception on ecosystem services as an adaptation strategy in urban stream recovery programs in the City of São Paulo, Brazil. Environ. Manage. (2021). https://doi.org/10.1007/s00267-021-01471-0

Goepel, K.D.: Implementation of an online software tool for the analytic hierarchy process (AHP-OS). Int. J. Analytic Hierarchy Process **10**, 469–487 (2018). https://doi.org/10.13033/ijahp.v10i3.590

Kapetas, L., Fenner, R.: Integrating blue-green and grey infrastructure through an adaptation pathways approach to surface water flooding. Philos. Trans. Royal Soc. A Math. Phys. Eng. Sci. **378** (2020). https://doi.org/10.1098/rsta.2019.0204

Kazakis, N., Kougias, I., Patsialis, T.: Assessment of flood hazard areas at a regional scale using an index-based approach and Analytical Hierarchy Process: application in Rhodope-Evros region, Greece. Sci. Total Environ. **538**, 555–563 (2015). https://doi.org/10.1016/j.scitotenv.2015.08.055

Kömüşçü, A.Ü., Çelik, S.: Analysis of the Marmara flood in Turkey, 7–10 September 2009: an assessment from hydrometeorological perspective. Nat. Hazards **66**, 781–808 (2013). https://doi.org/10.1007/s11069-012-0521-x

MGM, Turkish State Meteorological Service (2021). https://www.mgm.gov.tr. Accessed 26 May 2021

Öztürk, D.: Determination of flood vulnerability using GIS based multicriteria decision analysis methods-a case study: South Marmara Basin. Doctorate thesis, Yıldız Teknik Üniversitesi Fen Bilimleri Enstitüsü (2009). (in Turkish)

Pagano, A., Giordano, R., Vurro, M.: A decision support system based on AHP for ranking strategies to manage emergencies on drinking water supply systems. Water Resour. Manage **35**(2), 613–628 (2021). https://doi.org/10.1007/s11269-020-02741-y

Raja, N.B., Çiçek, I., Türkoğlu, N., Aydin, O., Kawasaki, A.: Landslide susceptibility mapping of the Sera River Basin using logistic regression model. Nat. Hazards **85**(3), 1323–1346 (2017). https://doi.org/10.1007/s11069-016-2591-7

URL1: https://bpmsg.com/ahp/ahp-hiergini.php?sc=Ejymyr. Accessed 27 Aug 2021

van Leeuwen, K., Sjerps, R.: Istanbul: the challenges of integrated water resources management in Europa's megacity. Environ. Dev. Sustain. **18**(1), 1–17 (2016). https://doi.org/10.1007/s10668-015-9636-z

Vasconcelos, V.V., Vaz, C.L., Follador, M., Alves, M.A.S.: Floodable cross-sectional area and slope to the nearest drainage as extensions of the HAND model: mapping flood susceptibility in the region of Lucas do Rio Verde, Mato Grosso State. Brazil. Rev. GeoAmazonia **5**(9), 3–25 (2017)

Woodward, D.E., Hawkins, R.H., Jiang, R., Hjelmfelt, A.T., Van Mullem, J.A., Quan, D.: Runoff curve number method: examination of the initial abstraction ratio. In: World Water and Environmental Resources Congress 2003, Philadelphia, Pennsylvania, United States, pp. 1–10 (2003). https://doi.org/10.1061/40685(2003)308

Urban and Peri-Urban Agriculture as a Nature-Based Solution to Support Food Supply, Health and Well-Being in Bucharest Metropolitan Area During the COVID-19 Pandemic

Ines Grigorescu[✉], Elena-Ana Popovici, Irena Mocanu, Mihaela Sima, Monica Dumitrașcu, Nicoleta Damian, Bianca Mitrică, and Cristina Dumitrică

Institute of Geography, Romanian Academy, 12, Dimitrie Racoviță Street, Sector 2, 023993 Bucharest, Romania

Abstract. Cities and metropolitan areas are becoming increasingly vulnerable to various environmental and socio-economic disturbances, posing a growing concern to food security, public health and well-being. This chapter intends to address two concepts that have only recently been addressed together – urban and peri-urban agriculture (UPUA) and nature-based solutions (NbS) – through their potential to mitigate the effects of the COVID-19 pandemic associated challenges in Bucharest Metropolitan Area. The work relies on statistical and spatial data related to UPUA as NbS initiatives, being identified two main types: bottom-up (e.g. community, private companies, or NGOs) and top-down (from the local authorities). Lessons learnt from experimenting with NBS during the pandemic period can be relevant for urban governance to further innovate and replicate the successful initiatives.

Keywords: Urban and peri-urban agriculture (UPUA) · Nature-based solutions (NbS) · Bucharest Metropolitan Area (BMA) · COVID-19 · Romania

1 Introduction

Cities resilience has been often challenged by various environmental, socio-economic and political drivers (e.g., extreme events, pollution, diseases) placing the current COVID-19 pandemic among them (Bayulken et al. 2020). This "new normal" situation of social distancing and travel restrictions has affected people's food security, health and well-being, especially within cities. During crisis periods, a weakened society should focus on improveing its well-being through the prioritisation of Nature-based Solutions (NbS) (UN News 2021). The concept was adopted by the European Commission in its Horizon 2020 research programme with an explicit focus on urban areas and urban regeneration by revitalizing the vacant space through community gardens and urban farms (EC 2015). NbS is an "umbrella concept" for other established nature-based approaches such as Green Infrastructure (GI) or ecosystem services (ESS) (Pauleit et al. 2017).

© The Author(s), under exclusive license to Springer Nature Switzerland AG 2022
D. La Rosa and R. Privitera (Eds.): INPUT 2021, LNCE 242, pp. 29–37, 2022.
https://doi.org/10.1007/978-3-030-96985-1_4

NbS are actions sustained by nature, which intend to strengthen community cohesion aimed at reconnecting people with nature (European Commission 2015). NbS provide valuable approaches for improving urban resilience and sustainability (Bayulken et al. 2020). However, most of the current research on NbS focuses on climate resilience in urban areas, followed by the protection of biodiversity and quality of life, while connecting it with UPUA and urbanization challenges is still unevenly addressed (Artmann and Sartison 2018). Although it encompasses multifunctional land use and food production in and around urban areas (from leisure to commercial activities) (Piorr et al. 2018), it has rarely been integrated into the urban green infrastructure (Rolf et al. 2020). Thus, relating urbanization and sustainable agriculture becomes a major societal challenge which will connect people with nature (Artmann and Sartison 2018). In addition, urban green space provides beneficial effects on physical and mental health, well-being and quality of life (van den Bosch and Sang 2017; Tsai et al. 2018) which are amplified in societal crisis (Ugolini et al. 2020).

Pandemic-related knowledge has shown the beneficial effects of UPUA as NbS in maintaining health and well-being (e.g., Martin et al. 2020; Soga et al. 2021; Ugolini et al. 2020; Theodorou et al. 2021) and enhancing pro-environmental behaviour (Martin et al. 2020). In Romania, addressing environmental and societal challenges through NbS is in early stage; only some contributions to specific challenges i.e., land fragmentation and grabbing (Petrescu-Mag et al. 2017) and solutions i.e., green (Gaiță 2018; Niță et al. 2018) and blue (Iojă et al. 2021) infrastructure in urban areas, were developed. This study aims to fill in this knowledge gap by stressing the role of UPUA as NbS during the COVID-19 pandemic in one of the most complex urban-rural systems in Romania: Bucharest Metropolitan Area (BMA).

2   Scope and Methodology

There were at least three major stages during the COVID-19 pandemic that significantly affected both urban dwellers and farmers, such as: *(1)* the total lockdown (between March 15–May 15, 2020) which imposed very strict measures (i.e. movement restrictions of residents, workers, farmers and goods), *(2)* the closing in November 2020 of the agri-food open-air markets and because of *(3)* the imposed quarantine (between March 22–April 25, 2021) to the localities surrounding the Capital-city, included the "vegetable pools" of Bucharest. The city and the farmers were caught unprepared. As a result, using of urban and sub-urban space to provide supply services for the population, as a part of a NbS scheme that can help addresses crisis situations should have been key during the COVID-19 pandemic.

This paper intends to approach two notions that have only recently been addressed together – UPUA and NbS – through the socio-ecological challenges BMA was exposed to during the COVID-19 crisis, yet hampered by the underlying conditions: the prevalent agricultural area, with major spatial and structural transformations after the fall of communism (i.e., land use/land cover changes, land fragmentation, cropland abandonment), urban sprawl, land concentration/grabbing and the increasing demand for fresh products and recreational spaces to supply the "core" of BMA – Bucharest, Romania's largest city with nearly 2 million inhabitants.

The authors used multiple information sources to highlight the UPUA's contribution as NbS to support resilience: data-oriented (statistical and spatial) and object-oriented (focused on UPUA initiatives as NbS). The data-oriented design used quantitative data (geospatial, statistical, published sources) available at Local Administrative Unit (LAU) level provided by/extracted from: National Institute of Statistics; Copernicus Land Monitoring (Fig. 1); Ministry of Agriculture and Rural Development to assess the general context of UPUA in BMA: land use/cover, functions and structure of green spaces, farm type, size and attributes. The object-oriented design relied on crowdsourcing, i.e., Facebook pages, community groups, online interviews, blogs, event pages, to identify relevant NbS-oriented projects that used UPUA to build resilience during the pandemic.

3   UPUA as NbS in Bucharest Metropolitan Area

UPUA in the study area is diversified and fragmented, having various functions from wellbeing, recreational, didactic, aesthetic or therapeutic to supply of fresh and healthy food to the urban population. During the COVID-19 lockdown and after, working from home has become the normal at a larger scale, many residents began to understand the need for green space and look for alternatives inside and outside the city (metropolitan area), by buying or renting a detached house with a yard.

Urban Agriculture. The specific of collective housing in the urban areas (i.e., Bucharest), built mostly in the communist period, consist of block of flats with limited outdoor space or with small balconies, mostly closed. Four types of urban agriculture have been identified: residential gardens, public gardens, abandoned or vacant land and rooftop gardens (Fig. 1, Table 1). *Residential gardens* are generally made up of communal gardens in the surrounding of block of flats and the detached houses gardens. *Public gardens* have the advantage of being easily integrated into initiatives carried out by local authorities or public institutions. Also, through concession or rental, they can be included into urban gardening-related projects carried out by private companies or NGOs. They generally support educational events, along with downshifting lifestyle-related activities. *Rooftop gardens* are limited, restricted to the city of Bucharest where the initiatives have started recently, mainly initiated by private companies. Thus, the rooftop gardens are placed on private investments: residential complexes (Stejarii Sports Complex), shopping malls (Mega Mall), hotels (Green Shop at Radisson Blu Hotel) etc. The only initiative on a public building is a garden on the roof of the General Directorate of Social Assistance of Bucharest, a project initiated by the NGO. *Abandoned or vacant land* generally resulting from abandoned industrial units or demolition sites, can become very versatile, having the potential to contribute to resource-efficient food production and revitalization through community gardens and urban farms (Artmann and Sartison 2018).

Peri-urban agriculture is characterised by three main categories of farms: small local traditional, organic and leisure and recreational (Fig. 1, Table 1). *Small local traditional farms* includes individual, peasant farms that operate as subsistence and semi-subsistence farms. In many cases, they supplement their income by selling the surplus products in the agri-food markets or from small contracts with supermarkets, Ho.Re.Ca

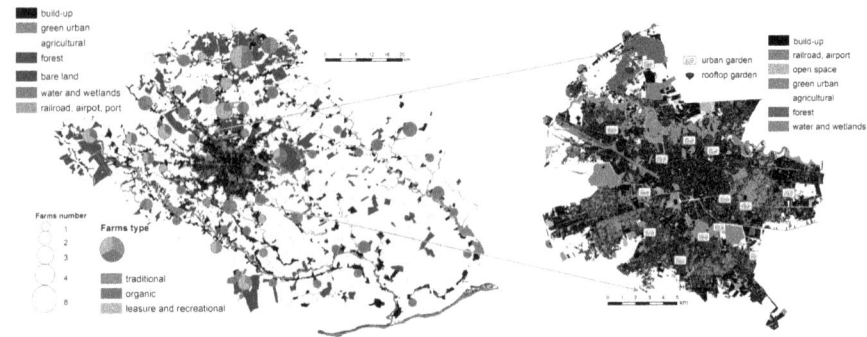

Fig. 1. Land use/land cover and main UPUA types in BMA (Sources: Urban Atlas, 2018, Street Tree Layer, 2018; Sentinel 2 Images, 2020)

Table 1. Main UPUA types as NbS in BMA

	Main UPUA types	Specialization/ potential as NbS	Products/ services/benefits	Size	Contribution to key societal challenges in urban areas
Urban	Residential gardens: detached house gardens; block gardens	recreational, aesthetics, gardening	fresh food supply, recreation, aesthetics	> 1 ha	biodiversity and ecosystem services, public health
	Public gardens	gardening, recreational, didactic, aesthetic, therapeutic	workshops, playgrounds, picking fruits/vegetables, horticultural therapy, animal feeding, crafts, camping, coach surfing, garden plot rental	> 1 ha	biodiversity and ecosystem services, public health
	Rooftop gardens	gardening, recreational, aesthetic, therapeutic, environmentally friendly	fresh food supply, recreation, aesthetics, air quality improvement, reducing urban heat island effect, water retention	> 1 ha	climate change, resource efficiency
	Abandoned or vacant land	recreational, gardening, didactic, aesthetic, therapeutic	urban renewal and regeneration, aesthetics	> 5 ha	urban renewal and regeneration
Peri-urban	Small local traditional farms	crop and animal production	vegetable and fruits, cheese and milk, vegetable and meats	> 5 ha	food security, economic growth
	Organic farms	crop production	vegetable and fruits	1 – 5 ha	food security, economic growth
	Leisure and recreational farms	leisure, recreational, didactic	picnic areas, wilderness, educational facilities, pick-your-own, harvest restaurants, overnight farm stays, on-farm direct sale, horseback riding	< 5 ha	public health, economic growth

Source: adapted after Artmann and Sartinson (2018); Sartinson and Artmann (2020) with authors' improvements

or school canteens, door-to-door, etc. This category was the most affected during the pandemic due to the imposed restrictions.

In addition to representing NbS through the ecosystem services they provide, *organic farms* proved to be more resilient during the pandemic due to the constant customers and the supply channels that operated almost unaffected by the restrictions (e.g., door-to-door, online shops). *Leisure and recreational farms* generally have a multi-functional character (Table 1); however, during the pandemic, the travel restrictions and the closure of the hospitality sector have led to the temporary or permanent shutdown of these farms, considered "non-essential sectors".

4 Building Resilience with UPUA-Related Initiatives as NbS in BMA: Examples of Pandemic Activities

Bottom-Up Adaptation Initiatives. These refer to individual, community initiatives (block associations, groups of friends, food groups), private companies or firms, usually supermarkets chains or gas stations (e.g., Kaufland, Lidl, MOL) or NGOs. The community initiatives are largely targeting the revitalization of green spaces between blocks by replacing the metal fences with hedges or planting flowers, vegetables or herbs in the green or vacant spaces between blocks. During the pandemic, these actions were extended to many blocks of flats by creating spaces for outdoor activities. Four types of initiatives were identified: *i) public-private partnerships* i.e., Lidl together with Technological College "Viaceslav Harnaj" (agricultural-technical college) which developed Băneasa Educational Farm; *ii) public-civil society partnerships* i.e., the Bucharest Botanical Garden together with Team Work NGO with project the Garden inside the Botanical Garden; *iii) private-civil society partnerships* i.e., Kaufalnd and Permaculture Research Institute NGO (10 project gardens "Grădinescu"), MOL Romania and the Partnership Foundation ("Green Spaces" project); *Grădinescu* was initiated by the Permaculture Research Institute, funded by Kaufland, and aims to create urban rooftop gardens, parking lots or behind shops. The gardens are designed to be cared for by the residents, with the supervision and guidance of the Permaculture Research Institute team. *iv) individual-civil society partnerships* which involved Urban Community Gardens NGO ("Gura Siriului Garden"). *Gura Siriului Garden* was developed the "School in the Garden" project and was launched in September 2020. Since schools and kindergartens were closed during most of the pandemic, this initiative was aimed at re-creating a space where children could meet, play and learn in a safe environment. The main activities unfolded are: gardening, workshops, composting, etc. A series of private initiatives to revitalize the short supply chains of some entrepreneurs and small local producers from the peri-urban area were also identified (online platforms, blogs and Facebook pages).

Top-Down Adaptation Initiatives. These initiatives started from the level of local authorities, such as town halls, with the role of using the residential green space around the blocks of flats as an agricultural space to gain an aesthetic, social or recreational function. They generally refer to the involvement of the authorities in the establishment of urban gardens in schoolyards for educational purposes, in the communal gardens of blocks of flats, on rooftops, abandoned land, city outskirts, etc.

The initiative of the City Hall of Sector 6, through the Public Domain and Urban Development Administration to implement an extensive program for the revitalization of the block gardens, related to the residential buildings. Within this initiative, the owners' associations that will enrol in the program will benefit from a series of facilities: landscaping, based on a tailored project made by professional landscapers; a voucher of maximum 2,500 lei per year (500-euro equivalent) for the payment of water-related expenses; dendro-floricultural materials; urban furniture etc.[1].

---

[1] https://adps6.ro/amplu-proiect-de-revitalizare-a-gradinilor-de-bloc-din-sectorul-6/.

The initiative of Ministry of Agriculture and Rural Development (MADR) to create a platform: https://www.rndr.ro/legume/, where small vegetable producers can add the surplus quantities of vegetables to reach the large retail chains.

The two types of initiatives should co-exist and be correlated to ensure the sustainability of future projects.

5   Discussions and Conclusions

The continuing urban development and metropolitan growth have increased the pressure on land resources. Apart from the mainstream spatial development, i.e. (sub)urbanization, that is currently affecting metropolitan areas, crises situations, such as the COVID-19 pandemic, have shown the vulnerability of cities (Pulighe and Lupia 2020; Vittuari et al. 2021).

The pandemic has significantly changed people's behaviour, leisure and travel options. Pre-existing problems faced by metropolitan areas related to leisure and social cohesion both in urban (limited green spaces, congestion of parks) and peri-urban (multi-functional farms) areas, as well as food security and safety have been enhanced by the current context. Despite the multiple negative effects, the COVID-19 pandemic also brought a series of opportunities by using creativity and innovation to revitalise activities and/or embrace new ones. This study showed that local authorities are not prepared or willing to adopt NbS by exploiting the potential of the local setting to meet the environmental and socio-economic needs of the community, especially under natural or human-induced disturbances. Most of the identified initiatives to use UPUA as NbS were those that started from citizens, NGOs or private companies, i.e., bottom-up. The increased interest in this type of activity, particularly observed during the pandemic when people felt the need to connect with nature, should encourage local authorities to get involved in creating urban gardens and support the development of common green spaces for the benefit of communities. The COVID-19 pandemic only amplified the discussion about the acute need for green spaces and outdoor activities in the urban environment and about the hidden potential of undervalued places that surrounds us (e.g., blocks of flats rooftops and gardens, courtyards). There are very few examples of well-established urban gardens based on community involvement. This have shown the critical role of community-supported agriculture during the pandemic (community-centred farms and gardens) in tackling food security, health and well-being. Overall, as a result of the identified initiatives, the pandemic experience has revealed the UPUA contribution as NbS to building COVID-19 resilience in terms of: food security & safety, health-related quality of life & well-being, social cohesion & inclusion, and civic involvement & awareness (Fig. 2). *Food security & safety* has been achieved by way of direct linking of producers with retailers and/or consumers through e-commerce and door-to-door delivery. In the study area it was found that UPUA types have differently contributed to this desideratum.

Food quality and human health are interlinked; however, during the pandemic, UPUA has also contributed to ensuring mental and physical health while coping with the social isolation and distancing through relaxing, socializing and gardening activities, having a particular role in providing *health-related quality of life & well-being*. Through social interaction, the latter contribution builds *social cohesion & inclusion* by providing equitable access of population to the benefits of UPUA while straightening well-being. Another essential contribution is *civic involvement & awareness* which stimulates public engagement in identifying and implementing ideas and initiatives for the benefit of the community.

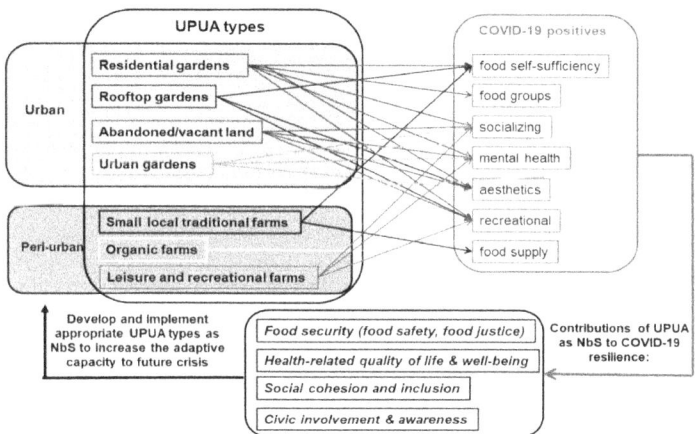

Fig. 2. UPUA contribution as NbS to COVID-19 resilience in BMA

If before the COVID-19 pandemic the concept of UPUA as NbS had only begun to take shape, during the restrictions, people have strived to find solutions to use any green space, abandoned, vacant or agricultural land able to bring benefits. Each of the UPUA types showed various ways of delivering vital benefits to the population, from social gatherings to growing vegetables and fruits. The need to spend time outdoors, to socialize and supply with fresh products has stimulated the creativity of local communities, NGOs or private companies in making the best of the available green and farmland to support NbS. Lessons learnt from experimenting with NbS during the pandemic period can be relevant for urban governance to further innovate and replicate the successful initiatives in other places. Urban and metropolitan planning policies should be adapted to recognize and support UPUA, acknowledging their significant role for ensuring food security and safety, health and well-being, especially in crises situations. Some key elements of resilience and adaptation to be considered in coping with (post-)COVID-19 challenges could intermingle the top-down and bottom-up instruments for the effective planning of available and abandoned/vacant land, provide support for the sustainable use of public gardens, extend rooftop gardening, reconnecting with local food, networking, digital development of small farmers, supporting small producers and short supply chains, multi-functionally, supporting/development of rural-urban relationships etc.

Acknowledgements. The current research was elaborated in the framework of the project entitled "The geographical study of man-environment relationships in metropolitan areas" – study made under the research plan of the Institute of Geography, Romanian Academy.

References

Artmann, M., Sartison, K.: The role of urban agriculture as a nature-based solution: a review for developing a systemic assessment framework. Sustainability **10**(6), 1937 (2018)

Bayulken, B., Huisingh, D., Fisher, P.M.: How are nature based solutions helping in the greening of cities in the context of crises such as climate change and pandemics? A comprehensive review. J. Clean. Prod 125569 (2020)

EC - European Commission: Towards an EU Research and Innovation Policy Agenda for Nature-Based Solutions & Re-Naturing Cities: Final Report of the Horizon 2020 Expert Group on 'nature-Based Solutions and Re-Naturing Cities'. EC, Brussels, Belgium (2015)

Gaiță, L.: The healing grid project: unlocking the potential of nature based solutions in Timisoara, Romania. In: IFoU 2018: Reframing Urban Resilience Implementation: Aligning Sustainability and Resilience, Climate Resilience Governance and Planning (2018)

Iojă, C.I., Badiu, D.L., Haase, D., Hossu, A.C., Niță, M.R.: How about water? Urban blue infrastructure management in Romania. Cities **110**, 103084 (2021)

Martin, L., White, M.P., Hunt, A., Richardson, M., Pahl, S., Burt, J.: Nature contact, nature connectedness and associations with health, wellbeing and pro-environmental behaviours. J. Environ. Psychol. **68**, 101389 (2020)

Niță, M.R., et al.: Using local knowledge and sustainable transport to promote a greener city: the case of Bucharest, Romania. Environ. Res. **160**, 331–338 (2018)

Pauleit, S., Zölch, T., Hansen, R., Randrup, T., Konijnendijk van den Bosch, C.: Nature-based solutions and climate change – four shades of green. In: Kabisch, N., Korn, H., Stadler, J., Bonn, A. (eds.) Nature-Based Solutions to Climate Change Adaptation in Urban Areas. TPUST, pp. 29–49. Springer, Cham (2017). https://doi.org/10.1007/978-3-319-56091-5_3

Petrescu-Mag, R.M., Petrescu, D.C., Petrescu-Mag, I.V.: Whereto land fragmentation–land grabbing in Romania? The place of negotiation in reaching win–win community-based solutions. Land Use Policy **64**, 174–185 (2017)

Piorr, A., Zasada, I., Doernberg, A., Zoll, F., Ramme, W.: Research for AGRI Committee – Urban and Peri-urban Agriculture in the EU, European Parliament, Policy Department for Structural and Cohesion Policies, Brussels (2018)

Pulighe, G., Lupia, F.: Food first: COVID-19 outbreak and cities lockdown a booster for a wider vision on urban agriculture. Sustainability **12**(12), 5012 (2020)

Rolf, W., Diehl, K., Zasada, I., Wiggering, H.: Integrating farmland in urban green infrastructure planning. An evidence synthesis for informed policymaking. Land Use Policy **99**, 104823 (2020)

Sartison, K., Artmann, M.: Edible cities–an innovative nature-based solution for urban sustainability transformation? An explorative study of urban food production in German cities. Urban Forestry Urban Greening **49**, 126604 (2020)

Soga, M., Evans, M.J., Tsuchiya, K., Fukano, Y.: A room with a green view: the importance of nearby nature for mental health during the COVID-19 pandemic. Ecol. Appl. **31**(2), e2248 (2021)

Theodorou, A., Panno, A., Carrus, G., Carbone, G.A., Massullo, C., Imperatori, C.: Stay home, stay safe, stay green: the role of gardening activities on mental health during the Covid-19 home confinement. Urban Forestry Urban Greening **61**, 127091 (2021)

Tsai, W.-L., et al.: Relationships between characteristics of urban green land cover and mental health in U.S. Metropolitan areas. Int. J. Environ. Res. Public Health. **15**, 340 (2018)

Ugolini, F., et al.: Effects of the COVID-19 pandemic on the use and perceptions of urban green space: an international exploratory study. Urban Forestry Urban Greening **56** (2020)

UN News: Critical year to 'reset our relationship with nature' (2021) van den Bosch, M., Sang, O.: Urban natural environments as nature-based solutions for improved public health – a systematic review of reviews. Environ. Res. **2017**(158), 373–384 (2017). https://doi.org/10.1016/j.envres.2017.05.040

Vittuari, M., et al.: Envisioning the future of European food systems: approaches and research priorities after COVID-19. Front. Sustain. Food Syst. **5**, 58 (2021)

Planning Green Spaces Investments for Improving Health and Well-Being in Cities Through Valuing Urban Nature

Riccardo Privitera[1]($\boxtimes$) and Jing Ma[2]

[1] Department of Civil Engineering and Architecture, University of Catania, Catania, Italy
riccardo.privitera@unict.it
[2] Department of Civil, Environmental and Natural Resources Engineering,
Luleå University of Technology, Luleå, Sweden
jing.ma@ltu.se

Abstract. Fundamental goals of urban green spaces investment and management are to provide natural features and man-made facilities and amenities that offer people satisfying leisure activities and nature experience by providing near-natural habitats and protecting biodiversity. The concept of human well-being is now attracting increasing attention in environmental science, policy, and management due to the inclusion of people and human societies in definitions of ecosystem. To improve human health and well-being, it requires that planners and policy makers generate a better understanding of the benefits provided by Ecosystem Services. A correct, explicit, and appropriate method for valuing natural and non-market goods and services is necessary. Within the current paper, the focus is to generate an overall understanding of identifying ways of measuring Ecosystem Services provided by urban nature in the context of Cost-Benefit Analysis. It looks at their different evaluation measures and focuses to address the limitations and the challenges.

Keywords: Urban nature · Ecosystem services and disservices · Human health and well-being · Cost-benefit analysis

1 Introduction

Urban green spaces are the places for nature experience for people within the cities. This nature for people or urban nature paradigm has been increasingly adopted by governments and non-profit organisations to frame, plan, and allocate resources (Posner et al. 2016). Urban nature promotes biodiversity conservation as well as the enhancement of the quality of life (Voigt and Wurster 2015). Fundamental goals of urban green spaces investment and management are to provide natural features and man-made facilities and amenities that offer people satisfying leisure activities and nature experience by providing near-natural habitats and protecting biodiversity. In the last two decades, a broad range of empirical research has begun detailing the effects of urban nature on people's physical, psychological, and spiritual well-being (Knecht 2004). The concept of human

© The Author(s), under exclusive license to Springer Nature Switzerland AG 2022
D. La Rosa and R. Privitera (Eds.): INPUT 2021, LNCE 242, pp. 38–46, 2022.
https://doi.org/10.1007/978-3-030-96985-1_5

well-being is now attracting increasing attention in environmental science, policy, and management due to the inclusion of people and human societies in definitions of ecosystem (Mace 2014). Human well-being evokes, variably, quality of life, happiness, and the social and economic conditions of individuals, communities and societies. Ecosystems in urban areas are the conditions and processes through which natural ecosystems and the species that compose them sustain and fulfil human life. They are directly accessed, enjoyed, consumed, or used to yield human well-being to the most pressing challenges for cities, from climate change adaptation and mitigation to citizens' health. Following the Millennium Ecosystem Assessment (MA 2005), the past decade has seen a growing interest in the impacts of ecosystem services on human well-being, with health as one of its main components. An extensive approach to health from the MA 2005) identifies that health relates to a feeling of strength, being nourished and able to access to adequate air and water (WHO CBD Secretariat 2015). Health is one of the most important factors for determining population well-being and depends on the conditions and functioning of the ecosystems and their ability to provide adequate and healthy flows of services (Martinez-Juarez et al. 2015).

To improve human health and well-being, it requires that planners and policy makers make a better understanding of the benefits provided by Ecosystem Services (ES) (Privitera and La Rosa 2018). This incorporates ES into decision making on the measure of how much a change in ecological conditions affects people, social benefit, or value to society (Olander et al. 2018). This research applies economic evaluation on urban nature investments, with the increasingly used Cost-Benefit Analysis (CBA). When a policy or a project is underway, normally it has two purposes of doing a CBA: one is to determine if a decision or an investment is feasible which means the benefits outweigh the costs and by how much; the other is to compare the total costs of each option against the total expected benefits (Atkinson and Mourato 2006). It is to demonstrate that the values of nature-based investment for improving public health and wellbeing have more benefits than costs.

Taking the far-reaching developments of the evaluation of urban nature into consideration, the aim of this paper is to identify key issues in any attempt to evaluate green spaces investments in a way that ES has been incorporated into a continuous and coherent CBA. Each way of measuring the value of urban nature from ES and CBA is distinctive. A lack of application in policy implementation with respect to take into account both measures for urban nature development and planning is explained by the less obvious and indirect benefits of urban nature. In order to make the public and related sectors aware of the significance of the connection between urban nature and their impacts in human health and well-being, it is necessary to review the ES provided by urban nature and its costs and benefits in a correct, explicit, and appropriate way. Within the current paper, the focus is to generate an overall understanding of identifying ways of measuring ES of urban nature in the context of CBA. It looks at their different evaluation measures and focuses to address the limitations and the challenges.

2 Valuing Ecosystem Services and Disservices

Research of ES seeks to catalogue the benefits that ecosystems provide to humans for their conservation and sustainable use (Evers et al. 2018). These assessments distinguish provisioning, regulating, cultural, and supporting ES and focus on the overlaps of and interdependencies between these categories (Bouwma et al. 2018). ES-health complex relationship, has been explored by (Oosterbroek et al. 2016) and described as the final health endpoints impacted by beneficial effects of ecosystems, using the same categories as the WHO's global burden of disease study. According to (Oosterbroek et al. 2016) framework, ES can provide a very wide array of health outcomes such as decreasing in heat strokes and exhaustions, cardiovascular diseases, sense organ diseases, respiratory diseases, but also decreasing on mental and behavioural disorders, anxiety and even reduction of incidence of neonatal conditions and malignant neoplasms.

Other frameworks, which systematically take into account the complex relationship between ecosystems and human health, contextualise negative ecological effects as part of ecosystem structures and functions (von Döhren and Haase 2015). As ES has been increasingly discussed, negative ecological effects have also been introduced as Ecosystem Disservices (EDS). EDS are described by (Byg et al. 2017) as functions or properties of ecosystems that cause effects that are perceived as harmful, unpleasant or unwanted, including pathogens of humans, live-stock and crops, and pest species which eat or damage crops. (von Döhren and Haase 2015) define EDS as 'negative ecological effects or impacts have been described as harmful consequences of ecological change or as deficient ES caused, for example, by the loss of biodiversity'. EDS are associated with a wider range of negative natural effects, including health effects, such as the spread of allergenic pollen and the presence of animals as disease vectors (Lyytimäki et al. 2008); economic effects, such as damage to infrastructure and maintenance costs for urban green structures (Dobbs et al. 2011); and socio-cultural effects, such as the discomfort caused by dark green spaces (Tzoulas et al. 2007) or the presence of animals or their droppings (Lyytimäki and Sipila 2009). The core question of the concept of EDS is not about highlighting them per se, but about putting both ES and EDS under a common assessment framework. This is required in order to establish a comprehensive overview of the net effects of ecosystem functions for human wellbeing.

Ecosystem functions are essential for the provision of ES but are not the same as services. If anyone benefits from a given element or process of an ecosystem, that element or process is not a service (Olander et al. 2018). Following the service cascade proposed by (Haines-Young and Potschin 2010), ecosystem functions are capacities of ecosystems (such as carbon sequestration) providing the ES (such as climate regulation). To describe and quantify the benefits deriving from ES/EDS, appropriate metrics and indicators must be selected. Variables which provide aggregated information based on the context, are selected for specific management purposes, with a value showing the difference between existing states and aspired target scenarios. Indicators of bio-physical/ecological properties that characterise urban ecosystems strictly depend on the availability of field data, geographical distribution, and spatial scale (Ma et al. 2021). To assess the state of the environment, in most cases not only a single indicator is needed but a set of indicators has to be carefully selected (Niemeijer and de Groot 2008).

Economic valuation of benefits from ES, as goods and services that are valuable for people (MA 2005), is complex and subjective. The (TEEB 2010) distinguishes ecological, social and economic benefits and values, highlighting that valuing ES and associated benefits is not straightforward: some people could value their income higher than their cultural identity, and may be willing to give up this identity for wealth (TEEB 2010), while different values could be attached to a particular benefit. Moreover, the diversity of beneficiaries contributes to add an extra layer of complexity and plurality to the idea of ES value (Small et al. 2017).

Particularly, the economic assessment of ES/EDS can explore different forms of utility provided by ecosystems to humans. Indeed, different money-based valuation methods can be used to assess both marketable (e.g., provisioning) and non-marketable (e.g. regulating) ES (Nikodinoska, et al. 2018). The latter, although vital for human well-being, are often overlooked by market-based valuations (Farber et al. 2002). ES generated by urban areas can be firstly assessed in biophysical terms and then valued in money units (Nikodinoska, et al. 2018; Ma et al. 2021).

3   Cost-Benefit Analysis for Planning Green Spaces Investments

Economic valuation of urban nature happens at a distinguished aspect, estimating the direct and indirect costs and benefits on the project or investment level. There is an increasing trend of using economic valuation for nature-related investments in Europe (Fuller et al. 2005). Policy makers want to understand whether these investments can have a positive economic return on both regional and community scale.

The costs of urban nature include costs of land acquisition; costs associated with design and construction, operation and maintenance; employment payroll; landscaping, opportunity costs associated with the loss of income that would have obtained for other uses (Maland 2012). Obviously, all these costs are various depending on specific features of the urban nature, for example the location, size, facilities, and surrounding environment. With different investment purposes, the maintenance costs, the density of trees and shrubs or the presence of flower beds vary widely and could increase the costs. Considering different categories of expenditure and previous studies, it is assumed that 75%–95% is the maintenance costs which constitute a major part of the total costs (Tempesta 2015). These costs could be calculated via market prices of the materials and labour resources.

Compared with costs, valuing benefits of urban nature is way more complicated. Three main categories of benefits have been recognised: economic, social and environmental benefits (Nowak and Dwyer 2007). The economic benefits revealed from green spaces investment is the real estate values. House price is influenced positively with the urban nature nearby, demonstrating that people's willingness to pay for higher valued properties in order to stay closer to the green space in the recognition of a better living environment. For social and environmental benefits, benefits include but not limited to aesthetic improvement, emotional and spiritual experiences, physical health, cooling, air quality and water regulation (Konijnendijk et al. 2013). Many researches have been done to put prices on these benefits (Pearce et al. 2006). Among various types of evaluation methods, two main categories have been used widely for the services generated by natural and semi-natural ecosystems: the revealed preference methods and stated preferences methods (Tempesta 2015).

Revealed preference methods, as summarised in Table 1, including Hedonic Price method, Travel Cost method, Averting Behaviour and Defensive Expenditures and Costs of Illness, are used for intangible non-market goods and services (Boyle 2003; Ma et al. 2021).

Table 1. An overview of four revealed preference methods (Source: modified by authors after Boyle 2003 as cited in Atkinson and Mourato 2006)

Method	Revealed behaviour	Conceptual framework	Application field
Hedonic pricing	Property purchase, job choice	Demand for varied goods	Property value and salary
Travel cost	Participation in recreation activity	Household production; complementary goods	Recreation and Leisure
Averting behaviour and defensive expenditure	Time costs; purchases to avoid harm	Household production; substitute goods	Health mortality and morbidity
Costs of illness	Expenditures to treat illness	Treatment costs	Health morbidity

Each of the methods contains its own pros and cons. For a specific project, suitable methods could be mixed to achieve the research purposes or prioritised methods might be selected to address the focus. The outlined revealed behaviours, framework and application show that in most cases, individual or household is the main concern. These approaches are based on actual decisions made by each person or household for their expenditures considering different situations. This group of methods provides actual data on how much people will pay to secure a service or a non-market good, which in principle is more reliable than only considering people's preferences.

The second category, the stated preference methods (e.g. Contingent Valuation) offers survey-based approaches to estimate the willingness to pay for changes in provision of non-market goods (Pearce et al. 2006, pp. 105). A random sample of people is asked directly to express their maximum willingness to pay for a hypothetical change in the provision or the improvement of certain public facility or service (Atkinson and Mourato 2006, pp. 106). From 1990s, this method has been applied to the valuation of environmental impacts in many researches in different countries, ranging widely from wastewater management, outdoor recreation, forest protection, air quality, visibility, waste management, sanitation improvements, biodiversity and health impacts (Markandya, 2016). The contingent valuation is one of a few methods able to capture all types of benefits, however, there remain concerns about its validity and reliability. As shown in Fig. 1, there is a general recognition that the valuation for non-market goods greatly relies on benefit transfer, which is to take a unit value of a non-market good estimated in an original or primary study and using this estimate - after some adjustment - to value benefits that arise when a new policy is implemented (Atkinson and Mourato 2006, pp 254).

This method is increasingly used in the literature as it shows high efficiency in reducing the time and costs of carrying on new primary evaluation studies and surveys in contingent evaluation methods.

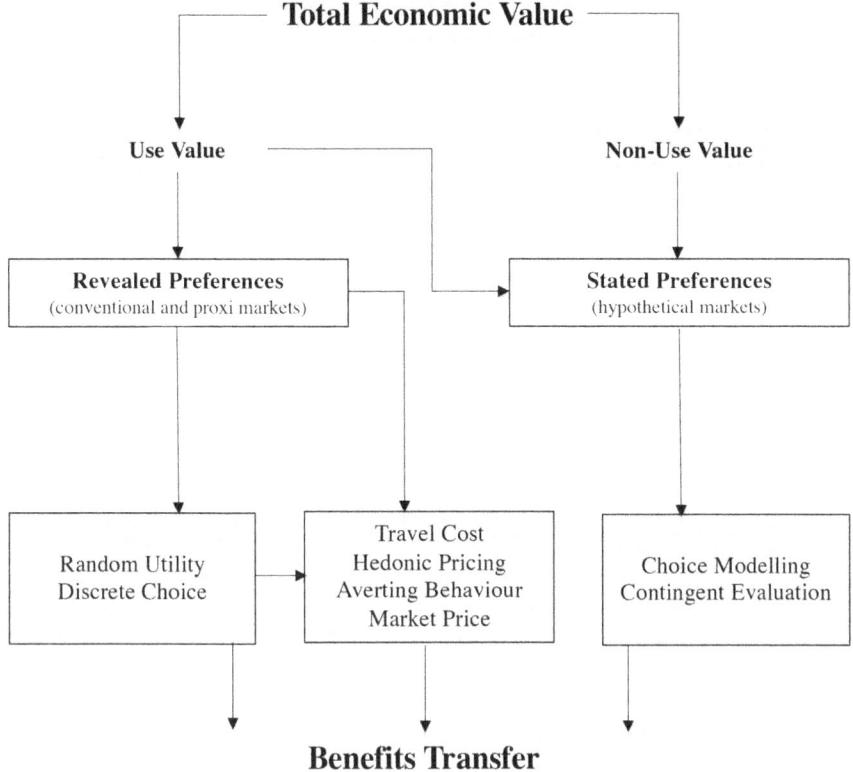

Fig. 1. Economic value and valuation techniques (Source: modified by authors after (DTLR 2002)).

4 Limitations and Challenges of Economic Valuation of Urban Nature

CBA is widely used to give guidance on the social evaluation of projects, investments and policies in monetary terms. For most economists, economic valuation and CBA are one and the same. When CBA is used for non-market goods, such as urban green spaces, the functions and the wider services which are not based on market price should be taken into account. It indicates that CBA is associated with individual preferences according to the fact that benefits are achieved on the basis of the individual utility of society's members. However, when evaluating ES according to the Millennium Ecosystem Assessment (MA 2005) a comprehensive assessment and the implications for human well-being are applied to provide a renewed scientific and reliable analysis. To provide a feasible nature-based economic assessment, it is necessary to facilitate a consensus about the applicability of ES and EDS to the CBA. Identifying the challenges and the limitations from these methods could bring an understanding of how to integrate ES/EDS when undertaking the CBA in the nature-based investment.

As emphasised in this research, a key challenge for planners and decision-makers is weighing up all relevant ES against the EDS that positively and negatively affect health outcomes, either directly or indirectly. According to (Oosterbroek et al. 2016), ecosystems providing a health service preventing one particular disease can at the same time provide a health disservice enhancing another disease. Whilst some ecosystems provide a health service concerning a particular disease, other ecosystems could provide a health disservice for the same disease. Each ES is not always possible to identify and associate a related EDS and vice versa. Within an ES/EDS-oriented CBA, where benefits are represented by ES while EDS constitute the costs, undertaking an economic valuation could imply an unbalanced assessment where costs and benefits would not be available at the same time for being compared and evaluated.

In the ES/EDS framework, metrics and indicators provide measures of ecosystem functions through the consideration of bio-physical features of the urban ecosystems. Each ecosystem function is related to an ES and then to a resulting benefit. Indicators indirectly measure ES/EDS and cannot provide any further information about the benefits/costs related to the services/disservices. Benefits and costs are not obvious and need to be appropriately identified according to the type of ES/EDS, the geographical distribution and the scale (local/regional/global) of the potential effects of the urban ecosystems, and the relevance for the economic and social context. The diversity of potential beneficiaries contributes to add an extra layer of complexity and plurality to the idea of ES benefits. The issue of selecting benefits is particularly relevant for the health-related outcomes whose relationship with ES is not so obvious and clear. Thus, a challenging stage of an ES/EDS-oriented CBA is the identification of appropriate benefits deriving from ES and costs deriving from EDS, that should be properly evaluated in monetary terms. Economic valuation of benefits from most of ES could be complex and subjective, especially for cultural ES and those other services providing non-material benefits. Closely linked with human emotional perceptions of the world, valuing these services in monetary terms could remain a real challenge.

In the context of measuring the value of urban nature, ES should be valued in terms of individual economic preferences. The question is if individuals may perceive intangible benefits from urban nature and if they may have difficulties to sufficiently understand ES and to decide what values of ES to integrate into a CBA. The traditional methods of CBA tend to use monetary units of measure and aggregate all values on a single scale. This way of measuring economic value has little use for measuring the value of ES especially when an ecosystem approach, an ecological threshold and certain ES are non-substitutable and scarce. In addition, the project or investment-based CBA calculates the net benefits based on a limited time frame. Within the preferred time period, it is possible to compare different solutions for achieving certain goals or to make the project feasible. This time frame differs when considering ES as services provided by urban nature which have no time limitation. For example, the benefits of reducing the urban heat island effect or increasing carbon sequestration could last as long as the urban nature exists. The dynamic ES with a longer life span increases the difficult of economic measure. Finally, economists are also clear that when they value an environmental asset, results are variously depending on the purposes and priorities of the investment project.

When the project is considering improving the amenities of the urban nature, a very small marginal change of ES might be excluded from the valuation.

While much of the investment in contemporary cities is on natural ecosystems, the fact is that whatever it is deliberately or inadvertently through pollution, the ecosystems are modified by human behaviour. The challenge for CBA is to apply certain kinds of measures that take economic and ecological values. The best way of undertaking a reasonable economic valuation on nature-based investment is to integrate the evaluation of ES/EDS into a CBA by adding additional criteria from the ES/EDS framework before taking a decision. This approach could help urban planners, developers and decision makers on achieving far better collaborative and agreed outcomes and planning more effective and sustainable green spaces investments for improving well-being and make cities healthier.

Acknowledgements. This paper is based on a research undertaken as part of the IWUN project funded by the Natural Environment Research Council, ESRC, BBSRC, AHRC and Defra [NERC grant reference number NE/N013565/1]. The authors are profoundly grateful to Prof. John Henneberry from the University of Sheffield (UK) for his invaluable support and guide on developing this research and for being such a great mentor.

This work has been also partially supported by the project ADDRESS under the University of Catania, Programme PIACERI Linea 2 2020–2022.

References

Atkinson, G., Mourato, S.: Cost-benefit analysis and the environment: recent developments. Organisation for Economic Co-operation and Development, Paris, France (2006)

Bouwma, I., et al.: Adoption of the ecosystem services concept in EU policies. Ecosyst. Serv. **29**, 213–222 (2018)

Boyle, K.J.: Introduction to Revealed Preference Methods. In: Champ, P.A., Boyle, K.J., Brown, T.C. (eds.) A Primer on Nonmarket Valuation, pp. 259–268. Kluwer, Dordrecht (2003)

Byg, A., et al.: Trees, soils, and warthogs – distribution of services and disservices from reforestation areas in southern Ethiopia. Forest Policy Econ. **84**, 112–119 (2017)

DTLR Department of Transport, Local Government and the Regions: Green Spaces, Better Places - The Final Report of the Urban Green Space Task Force. DTLR, London (2002)

Dobbs, C., Escobedo, F., Zipperer, W.: A framework for developing urban forest ecosystem services and goods indicators. Landsc. Urban Plan. **99**, 196–206 (2011)

Evers, C.R., et al.: The ecosystem services and biodiversity of novel ecosystems: a literature review. Global Eco. Conservation **13**, e00362 (2018)

Farber, S., Costanza, R., Wilson, M.: Economic and ecological concepts for valuing ecosystem services. Ecol. Econ. **41**, 375–392 (2002)

Fuller, K., Monson, M., Ward, J., Mathews, L.G.: Can nature drive economic growth? Rev. Agric. Econ. **27**(4), 621–629 (2005)

Haines-Young, R, Potschin, M.: The Links between Biodiversity, Ecosystem Services and Human Well-Being. In: Raffaelli, D., Frid, C. (eds.) Ecosystem Ecology: A New Synthesis, CUP, Cambridge (2010)

Knecht, C.: Urban nature and well-being: some empirical support and design implications. Berkeley Plan. J. **17**, 82–108 (2004)

Konijnendijk, C.C., Annerstedt, M., Nielsen, A.B., Maruthaveeran, S.: Benefits of urban parks: a systematic review. A report for IPFRA, IFPRA (2013)

Lyytimäki, J., Petersen, L.K., Normander, B., Bezák, P.: Nature as a nuisance? Ecosystem services and disservices to urban life style. Environ. Sci. **5**(3), 161–172 (2008)

Lyytimäki, J., Sipila, M.: Hopping on one leg – The challenge of ecosystem disservices for urban green management. Urban Forestry Urban Green. **8**, 309–315 (2009)

Ma, J., Henneberry, J., Privitera, R.: The challenges of valuing urban nature: accounting for urban ecosystem services within the framework of a cost-benefit analysis of nature-based investments. In: La Rosa, D., Privitera, R. (eds.) INPUT 2021. LNCE, vol. 146, pp. 81–90. Springer, Cham (2021). https://doi.org/10.1007/978-3-030-68824-0_9

Mace, G.M.: Whose conservation? Science **345**, 1558–1560 (2014)

Markandya, A.: Cost benefit analysis and the environment: How to best cover impacts on biodiversity and ecosystem services 101, OECD Publishing (2016)

Maland, M.M.: The return on investment of parks and open space. Doctoral dissertation, University of Georgia (USA) (2012)

Martinez-Juarez, P., Chiabai, A., Taylor, T., Gómez, S.Q.: The impact of ecosystems on human health and wellbeing: a critical review. J. Outdoor Recreat. Tour. **10**, 63–69 (2015)

Millennium Ecosystem Assessment (MA): Ecosystems and Human Wellbeing: A Framework for Assessment. Island Press, Washington DC (2005)

Niemeijer, D., de Groot, R.S.: Framing environmental indicators: moving from causal chains to causal networks. Environ. Dev. Sustain. **10**(1), 89–106 (2008)

Nikodinoska, N., Paletto, A., Pastorella, F., Granvik, M., Franzese, P.P.: Assessing, valuing and mapping ecosystem services at city level: the case of Uppsala (Sweden). Ecol. Model. **368**, 411–424 (2018)

Nowak, D.J., Dwyer, J.F.: Understanding the benefits and costs of urban forest ecosystems. In: Kuser, J.E. (eds) Urban and Community Forestry in the Northeast. Springer, Dordrecht (2007). https://doi.org/10.1007/978-1-4020-4289-8_2

Olander, L.P., et al.: Benefit relevant indicators: ecosystem services measures that link ecological and social outcomes. Ecol. Ind. **85**, 1262–1272 (2018)

Oosterbroek, B., de Kraker, J., Huynen, M.M.T.E., Martens, P.: Assessing ecosystem impacts on health: a tool review. Ecosyst. Serv. **17**, 237–254 (2016)

Pearce, D.W., Atkinson, G., Mourato, S.: Cost-benefit Analysis and the Environment: Recent Developments. OECD, Paris (2006)

Posner, S.M., McKenzie, E., Ricketts, T.H.: Policy impacts of ecosystem services knowledge. Proc. Natl. Acad. Sci. Unit. States Am. **113**(7), 1760–1765 (2016)

Privitera, R., La Rosa, D.: Reducing seismic vulnerability and energy demand of cities through green Infrastructure. Sustainability **10**(2591), 1–21 (2018)

Small, N., Munday, M., Durance, I.: The challenge of valuing ecosystem services that have no material benefits. Glob. Environ. Chang. **44**, 57–67 (2017)

Kumar, P.: TEEB: The Economics of Ecosystems and Biodiversity Ecological and Economic Foundations. In: Kumar, P. (ed.). Earthscan, London and Washington (2010)

Tempesta, T.: Benefits and costs of urban parks: a review. Aestimum **67**, 127 (2015)

Tzoulas, K., et al.: Promoting ecosystem and human health in urban areas using green infrastructure: a literature review. Landsc. Urban Plan. **81**(3), 167–178 (2007)

Voigt, A., Wurster, D.: Does diversity matter? The experience of urban nature's diversity: case study and cultural concept. Ecosyst. Serv. **12**, 200–208 (2015)

von Döhren, P., Haase, D.: Ecosystem disservices research: a review of the state of the art with a focus on cities. Ecol. Ind. **52**, 490–497 (2015)

WHO: CBD Secretariat Connecting Global Priorities: Biodiversity and Human Health: a State of Knowledge Review: World Health Organization (2015)

Open Spaces and Green Infrastructure – A Comparison with Planning Indications for the City of Rome

Daniele La Rosa[✉] [ⒾD]

Department Civil Engineering and Architecture, University of Catania, Catania, Italy
dlarosa@darc.unict.it

Abstract. Urban Green Infrastructure is increasingly acknowledged as a consolidated planning approach to ensure the provision of ecosystem services in cities and therefore increase their levels of resilience to different internal and external drivers and pressures. This paper argues that a fundamental step in the planning process of the Green Infrastructure must include the identification, analysis and mapping of all different types of open spaces that can be integrated and connected in a GI. In two urban-rural transects in the city of Rome, all different open spaces have been identified and mapped and then compared to the existing planning indications included in the current land use plan of the city. Results showed significant differences between the formally planned and recognized urban green space with the current open spaces and new opportunities to increase the size and quality of Green Infrastructure have been proposed.

Keywords: Open spaces · Green infrastructure · Land use plan · Rome

1 Green Infrastructure to Enhance Urban Resilience

Contemporary research on urban systems is looking at the concept of urban resilience as a new paradigm to achieve higher level of sustainability in cities that have to face future challenges and perturbations (Garcia and Vale 2017). Starting from ecological science, resilience has been defined by Holling (1973) as the "measure of the persistence of systems and of their ability to absorb change and disturbance and still maintain the same relationships between populations or state variables".

Along with other ecological concepts that have been transferred to from ecology to different scientific fields and application (i.e. as urban design and planning), resilience has taken upon a flexibility that is beyond its intended meaning (Gomes Ribeiro and Jardim Gonçalves 2019). For urban systems, the concept has been considered as the ability of a city to adapt and adjust to changing internal or external processes and rapidly return to desired functions in the face of a disturbance (Picket et al. 2004). Resilient cities, infrastructure, ecosystems and communities recover from natural and human-made hazards quickly and more effectively.

Urban resilience represents an interesting way to manage risk trough spatial planning, for example when looking at management of water in cities (White 2010; Eaton 2018),

© The Author(s), under exclusive license to Springer Nature Switzerland AG 2022
D. La Rosa and R. Privitera (Eds.): INPUT 2021, LNCE 242, pp. 47–55, 2022.
https://doi.org/10.1007/978-3-030-96985-1_6

pressures of urban development processes on (semi)natural ecosystems (Zhang 2016) and social and economic challenges derived by globalization (Saja et al. 2019). In this perspective resilience represents a relevant theoretical but also practical reference for urban planning and design that aims to minimize the impacts of urban transformations and developments (Shamsuddin 2020). Any planning policy and strategy should deal with the driving forces (on which the thresholds of the system depend) through action that reduce the hazard (mitigation actions) and exposure and vulnerability (adaptation actions) (La Rosa and Pappalardo 2020).

A fundamental role in ensuring and increase the levels of resilience in cities is played by the Green Infrastructure (GI), usually defined as a framework of connected ecological components that are able to deliver a wide assets of Ecosystem Services (La Rosa and Pappalardo 2021) and improve the overall quality of urban environments. GI includes different types of areas and recently such spaces have been increasingly recognised and discussed despite often being overlooked and, more importantly, ill-acknowledged by spatial planning (La Rosa 2019) as areas able to deliver multifunctional benefits for urban systems and their residents and therefore to increase urban resilience (Calderón-Contreras and Quiroz-Rosas 2017).

This paper argues that the first step in planning an effective GI is the identification and analysis of the all different types of urban open space that can be included into city planning and governance of GI. To this end, this research identifies and classifies the different types of urban open space that can be components of an urban GI. In a second step these components are compared to the current planning instrument currently active in the studied area, the city of Rome. Particularly, the current planning indications included in the land use plan of Rome are compared with the current conditions of the components of the GI.

2 Materials and Methods

2.1 Land Use and Planning Analysis with Random Transects

A wide range of existing spaces that can be potentially used as new elements of a GI exists in Rome, although not all of them are currently managed and/or used by residents. Possible components of the GI includes established and recognisable parks, squares and other open spaces but also a range of potential and less formally acknowledged open space types as terrain vague, non-urbanised areas, abandoned agricultural, industrial lands and other land-uses.

To address the large size of the entire Rome Metropolitan area, with the enormous task of compiling, quantifying and classifying all GI elements, transects have been used to provide a snapshot of the city complexity.

Transects have been used successfully to investigate diversity, complexity and temporal change in ecological studies but also extended recently to cities (Follmann et al. 2018). The transects have been generated by the Random Transect Generator (Geographic Business Solutions 2018), an add-in tool for ArcMAP that creates randomly located custom line transects within an area specified – in this case given by the extent of the metropolitan area of Rome (Fig. 1).

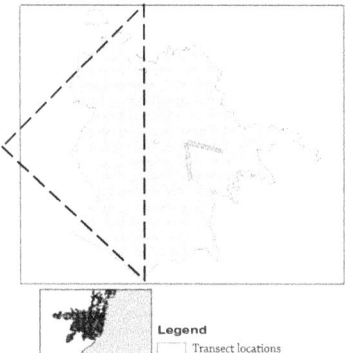

Fig. 1. The locations of their transects in Rome

2.2 Available Data

The metropolitan area of Rome includes 121 municipalities, with 4.3 million people living in over 5,363 km². The capital city covers an area of 1,287 km² with a population of about 2.9 million people. Valuable natural and semi-natural features and archaeological elements are present in the urban area and are safeguarded by a different types of protected areas, such as urban and urban parks. Land-use data used in this work has been obtained from the Urban Atlas land-use layer (European Environment Agency 2012) at an average scale of detail of 1:12,000 and based on SPOT 5 satellite imagery (with a resolution of 2.5 m). Urban Atlas includes the following classes of open spaces: green urban areas, sports and leisure facilities, arable land (annual crops), pastures, complex and mixed cultivation patterns, permanent crops, and land without current use. Information about planning was taken from the recent Land Use Masterplan (Città di Roma 2018) which included the following planning indications for the open spaces: Protected areas, Private protected green space, Local services, Urban services, Historical residences, Urban Rehabilitation and New developments.

3 Results

The two transects explored in Rome included areas from districts close to the urban centre to the rural sectors of the city (Figs. 2). They present a similar share of open spaces versus built up areas: in Transect 1 open spaces comprise 40%, while in Transect 2 they cover 28%.

Existing categories of open spaces across Transect 1 are mapped in Fig. 3 (up) and summarized in Fig. 4 (left). In this, the most prevalent categories of open space found were *green urban areas* (147 ha, 15% of the transect area), *sport and leisure facilities* (87 ha, 9%). The other present categories are *arable lands*, *pastures*, *mixed cultivation* and *permanent crops*, summarizing a total of 155 ha (16%). The most of open spaces area found in a peri-urban park, Parco dell'Appia Antica, which includes historical and archaeological sites in a matrix of farmlands and pastures.

Analysis of the planning indication for existing opens spaces in Transect 1 (Fig. 3, bottom) confirmed that most of these areas belonged to the category *Protected Areas*

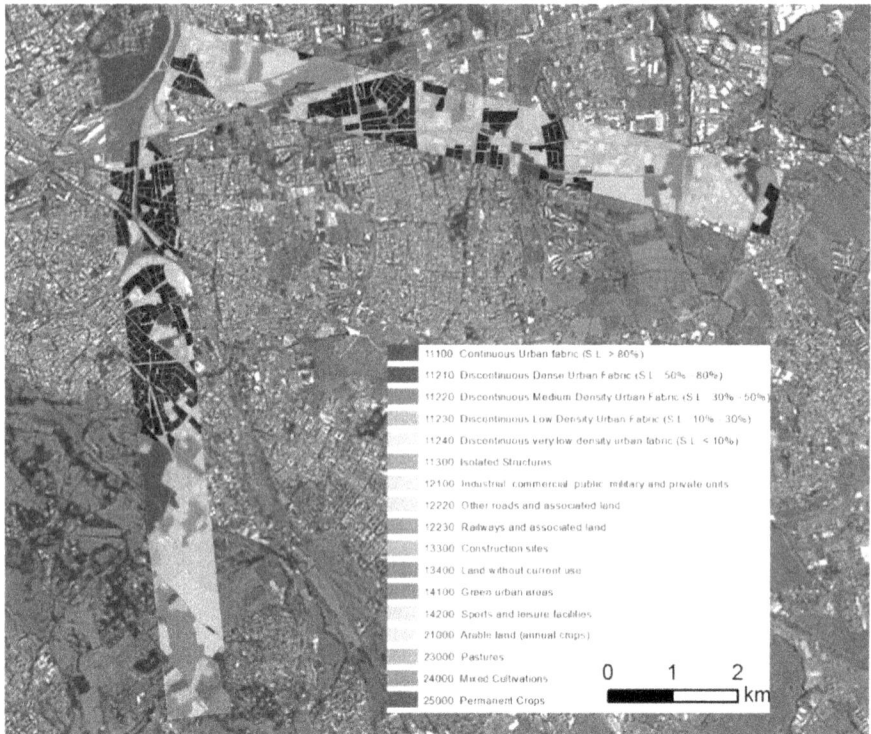

Fig. 2. Land-use classes in the transects

(305 ha, 77%), which covered the southern section of the transect. The 2nd and 3rd most recurring categories were *urban services* (62 ha, 16%) and *local services* (14 ha, 4%), which included small neighbouring parks or public/private sport facilities (Fig. 4). *Historical residences* (6 ha, 2%) and *Private protected green space* (3 ha, 1%) included few patches only.

Table 1 shows the distribution of all the categories of open spaces with the relative planning categories. *Green urban areas* represent the land-use corresponding to the highest number of planning categories (*Protected areas, Private protected green space, Local services, Urban services, Historical residences*) and this indicates the good potential and versatility of such areas for different planning objectives. Among the planning categories, *Protected areas* are the most frequent with 28 patches and include almost all the land-use categories found in the transect 1.

In Transect 2, the main categories of open spaces were *pastures* (89 ha, 9% of the transect area), *green urban areas* (49 ha, 7%) and *arable land* (38 ha, 4%). Other less frequent categories are *Construction sites* (10 ha, 2%), *sports and leisure facilities* (8 ha, 2%) and *land without current use* (7 ha, 1%). Land-use categories are mapped in Fig. 4, up.

The planning categories in Transect 2 (Fig. 4, bottom) were quite different from the ones found in Transect 1. The most frequent category – not present in Transect 1

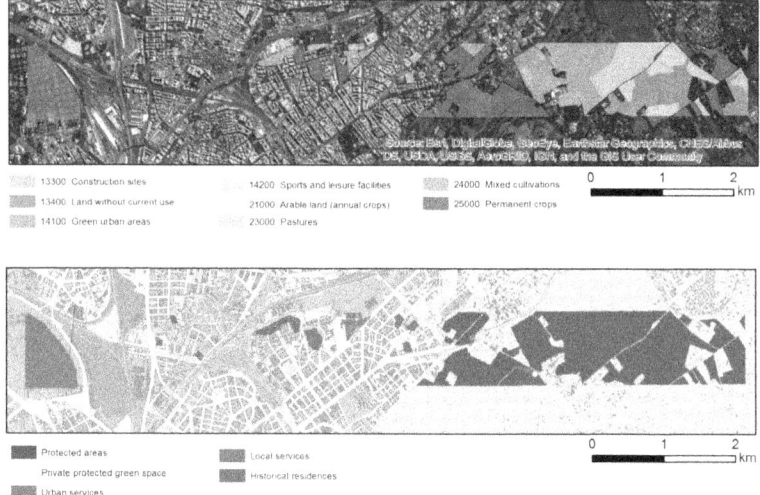

Fig. 3. Rome transect 1 - land use (up) and planning (bottom)

Table 1. Transect 1 – comparison between land-use and planning categories

Land-Use category	Planning category	# patches	Area (m2)
Land without current use	Protected areas	2	36368
Green urban areas	Protected areas	7	732264
Green urban areas	Private protected green space	1	37273
Green urban areas	Local services	3	95456
Green urban areas	Urban services	2	545132
Green urban areas	Private protected green space	1	2435
Green urban areas	Historical residences	3	63443
Sports and leisure facilities	Protected areas	4	729653
Sports and leisure facilities	Local services	4	47143
Sports and leisure facilities	Urban services	4	73597
Sports and leisure facilities	Private protected green space	2	19640
Arable land (annual crops)	Protected areas	7	730346
Pastures	Protected areas	4	269930
Mixed cultivations	Protected areas	1	463813
Permanent Crops	Protected areas	1	90208
			2180998

- was *Open Spaces (unmanaged)* (127 ha, 46% of the total area of open spaces). This very general category includes farmlands, abandoned farmlands and other rural areas with some vegetation. For this categories no specific indications or prescription are given in the urban plan of Rome. *Local* services (84 ha, 31%) and *Urban services* (23 ha, 8%) were also included in both Transects. An important planning category is *New development* (39 ha, 14%), which includes areas for which the planning foresees new urban development that will cancel or strongly modify some of the current open space and thereby decrease their potential of being included in a GI.

Table 2 reports how the different land-use categories of the open spaces are planned in Transect 2. *Arable land* is the land-use to which corresponds the higher number of planning categories (*Local services, Open spaces (unmanaged), New development, Rehabilitation*). When looking at the planning categories, open spaces represent a very general category including an high number of patches (33) of all the categories of open spaces found in the Transect 2 (*Construction sites, Land without current use, Green urban areas, Sports and leisure facilities, Arable land, Pastures*).

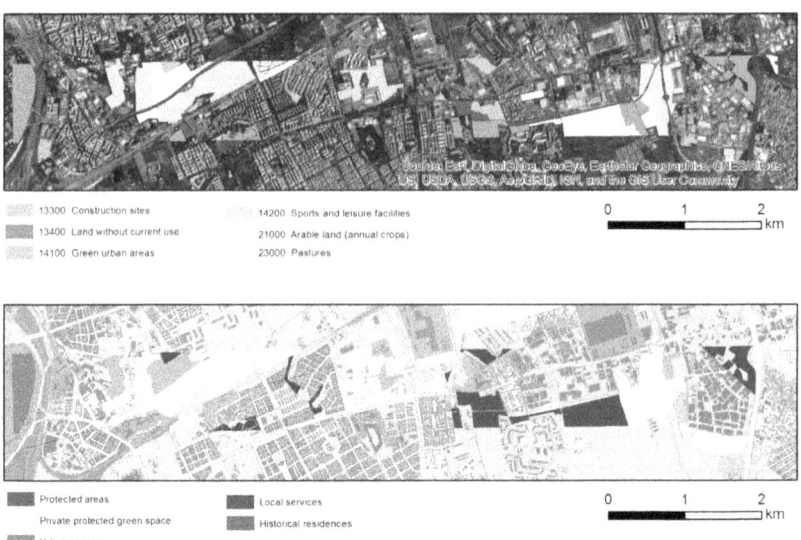

Fig. 4. Rome transect 2 – land use (up) and planning (bottom)

Table 2. Transect 2 – comparison between land-use and planning categories

Land Use	Planning category	# patches	Area (m2)
Construction sites	Open spaces (unmanaged)	2	120899
Construction sites	New development	4	57088
Land without current use	Local services	2	59693
Land without current use	Open spaces (unmanaged)	6	74093
Green urban areas	Local services	16	366368
Green urban areas	Urban services	1	229919
Green urban areas	Open spaces (unmanaged)	3	121736
Sports and leisure facilities	Local services	5	143884
Sports and leisure facilities	Open spaces (unmanaged)	1	8544
Arable land (annual crops)	Rehabilitation	1	24389
Arable land (annual crops)	Local services	2	275166
Arable land (annual crops)	Open spaces (unmanaged)	11	196796
Arable land (annual crops)	New development	1	184347
Pastures	Open spaces (unmanaged)	10	745541
Pastures	New development	2	146135

4 Discussions and Conclusions

The analysis of the two transects revealed a heterogeneity of land-uses or open space types which is mainly due to the length of each transect, which runs from the (almost) center to the peri-urban and rural part of the metropolitan area.

The comparison with the land use plan shows that only for a limited number of instances did planning indications match the characteristics of current open spaces. This is a first indicator that potentialities of open spaces (for example in terms of delivering urban ecosystem services) are not fully captured by planning.

Although the number of categories of open spaces were similar to those identified in planning, their characteristics were actually very different. For example, open spaces under the category of green urban areas had different planning categories, including protected areas, Private protected green space, Local public services, Urban services, Private protected green space, and Historical residences. At the same time, many unmanaged open spaces have been planned with a generic indication (i.e. open space unmanaged). For these areas the new planning indication can be foreseen, for example:

- peri-urban parks (large highly natural/semi-natural areas with relevant vegetation cover);
- different forms of urban agriculture (i.e. urban farms, allotments gardens and community supported agriculture);
- open space in educational services and business parks (areas with expansive open space including tree cover, remnant vegetation and grass);
- terrain vague (abandoned, natural or semi-natural lands within, around and between developed patches, mainly subject to spontaneous re-forestation through undisturbed succession);
- informal recreation areas (green spaces available for public access and enjoyment, but with limited management and only limited number of facilities);

The following general challenges can be highlighted to fully integrate open spaces in new and enhanced GI.

- difficulties/inertia by planning in incorporating and enabling potentialities of open spaces in a comprehensive GI;
- lack of public financial resources to purchase private areas to be included in a public GI due;
- the need for multi-stakeholder participation and action in the planning and management of the GI, especially for those green spaces that are owned or managed by private bodies, NGO or other local associations;
- the opportunity offered by strategic, multi-level spatial governance to implement GI as opposed to the varying planning objectives that are put forward by the many different municipalities that belong to one metropolitan area.

Finally, some limitation of this work should be highlighted. The different resolution and between the land use and planning geographical data generated some issues in terms of data accuracy: land-use dataset had higher resolution while planning data held simple polygons as broader, less detailed data. For example, in Rome the land-use data was from 2012 and the urban plan was from 2018. Developments that occurred later than these dates (for example in Transect 2) could have produced differences between land use asset and planning indications. To try to address these issues, the land-use data has been revised by the use of up-to-date aerial photograph and Google Map imagery and street views, so that for some instances categories of land uses were changed according the visual interpretation of Google imagery.

Overall the results obtained support the design of a multifunctional and inclusive system and network of abiotic, biotic and cultural elements (Ahern 2007). Achieving and actually implementing the GI requires a full understanding of the different socio ecological aspects of the existing categories open spaces so to provide to the municipalities – and other administrations in charge of planning – a range of new possibilities and differentiated planning options aimed at enlarging the size and connectivity of the GI and increasing important urban ecosystem services.

Acknowledgements. The present work has been partially supported under the PIA.CE.RI. research program of the University of Catania, 2020–2022, Project ADDRESS "Advanced Design for Demand Responsive transport and Services of general interest in inner areas for the Sustainable territorial re-equilibrium".

References

Ahern, J.: Green infrastructure for cities: the spatial dimension. In: Novotny, V., Brown, P. (eds.) Cities of the Future Towards Integrated Sustainable Water and Landscape Management. IWA Publishing, London (2007)

Calderón-Contreras, R., Quiroz-Rosas, L.E.: Analysing scale, quality and diversity of green infrastructure and the provision of Urban Ecosystem Services: a case from Mexico City. Ecosyst. Serv. **23**, 127–137 (2017). https://doi.org/10.1016/j.ecoser.2016.12.004

Città di Roma: Piano Regolatore Genearale – Sistemi e regole. http://www.urbanistica.comune.roma.it/prg-2008-vigente-disegno-definitivo.html

Eaton, T.T.: Approach and case-study of green infrastructure screening analysis for urban stormwater control. J. Environ. Manag. **209**, 495–504 (2018)

Follmann, A., Hartmann, G., Dannenberg, P.: Multi-temporal transect analysis of peri-urban developments in Faridabad, India. J. Maps **14**(1), 17–25 (2018)

Garcia, E.J., Vale, B.: Unravelling Sustainability and Resilience in the Built Environment. Routledge, London (2017)

Geographic Business Solutions: Random transect generator (2018)

Gomes Ribeiro, P.J., Jardim Gonçalves, L.A.P.: Urban resilience: a conceptual framework. Sustain. Cities Soc. **50**, 101625 (2019). https://doi.org/10.1016/j.scs.2019.101625

Holling, C.S.: Resilience and stability of ecological systems. IIASA, Laxenburg (1973)

La Rosa, D.: Why is the inclusion of the ecosystem services concept in urban planning so limited? A knowledge implementation and impact analysis of the Italian urban plans. Soc.-Ecol. Pract. Res. **1**(2), 83–91 (2019). https://doi.org/10.1007/s42532-019-00016-4

La Rosa, D., Pappalardo, V.: Planning for spatial equity - a performance based approach for sustainable urban drainage systems. Sustain. Cities Soc. **53**, 101885 (2020). https://doi.org/10.1016/j.scs.2019.101885

La Rosa, D., Pappalardo, V.: Policies and planning of urban green infrastructure and sustainable urban drainage systems. In: Catalano, C., Andreucci, M.B., Guarino, R., Bretzel, F., Leone, M., Pasta, S. (eds.) Urban Services to Ecosystems. FC, vol. 17, pp. 297–316. Springer, Cham (2021). https://doi.org/10.1007/978-3-030-75929-2_16

Pickett, S.T.A., Cadenasso, L.M., Grove, J.M.: Resilient cities: meaning, models and metaphor for integrating the ecologicalm, socio-economic, and planning realms. Landsc. Urban Plan. **69**, 369–384 (2004)

Saja, A.M.A., Goonetilleke, A., Teo, M., Ziyath, A.M.: A critical review of social resilience assessment frameworks in disaster management. Int. J. Disaster Risk Reduct. **35**, 101096 (2019). https://doi.org/10.1016/j.ijdrr.2019.101096

Shamsuddin, S.: Resilience resistance: the challenges and implications of urban resilience implementation. Cities **103**, 102763 (2020). https://doi.org/10.1016/j.cities.2020.102763

White, I.: Water and the City – Risk, Resilience and Planning for a Sustainable Future. The Natural and Built Environment Series. Routledge, London and New York (2010)

Zhang, X.Q.: The trends, promises and challenges of urbanisation in the world. Habitat Int. **54**, 241–252 (2016). https://doi.org/10.1016/j.habitatint.2015.11.018

The Architectural and Landscape Project for a Sustainable Urban Context. Use and Application of Nature-Based Solutions to Manage Urban Floods

Salvatore Leanza(✉)

Adrano, Italy
ing.salvo.leanza@gmail.com

Abstract. Historically, every human being has a strong link with water as a natural element. But compared to the past, since human activities began to influence climate changes, water has been converted into a dangerous, violent, and often mortal element. Water, in fact, arrives in the cities without having the possibility of flowing in the natural ground since we have cemented our spaces. For this reason, in the last years, we have had to cope with more severe urban floods and find new solutions for a better urban life. The most widespread paradigm is to allow water to enter our cities and convert them into sponge cities. To do this, using nature-based solutions in the new urban project is essential. Thanks to the analysis of different case studies from various European regions, such as the Nordic regions, Netherlands and Spain, this paper aims to highlight how integrated use of nature-based solutions in the contemporaneous architectural and urban projects is essential if we want to survive future urban floods. At the same time, it aims to demonstrate that the use of NBS in contemporary urban and landscape projects represents the opportunity to modify our lifestyle, spaces, and relation with nature, improving our way of living, from the environmental, social, and economic points of view. In conclusion, the paper aims to ask some questions for future cities and try to open a debate on the future of the urban contexts.

Keywords: Design with water · Climate change adaptation · Public spaces · Sustainable design

1 Introduction

Every human being always has had a strong bond with water as a natural element indispensable for his life. For this reason, in the 7th century BC, the Greek philosopher Thales of Miletus identified the water as the *arche*, namely the beginning of everything, which influences men life and culture. We are 70% water and grow in the maternal uterus immersed in the amniotic liquid being 99% water. So, even if we consider ourselves "terrestrial animals", we could say we come from water intended as a lulling element with feminine and maternal characteristics (Bachelard 1942).

S. Leanza——Freelance Building Engineer.

© The Author(s), under exclusive license to Springer Nature Switzerland AG 2022
D. La Rosa and R. Privitera (Eds.): INPUT 2021, LNCE 242, pp. 56–63, 2022.
https://doi.org/10.1007/978-3-030-96985-1_7

Although water has not an own shape, it continually moulds our planet. It is the only natural element that puts in relation all Earth ecosystems to each other. Thus, connected through the hydrogeological and hydrological cycles, Earth ecosystems can be considered as a wider, single, closed and balanced system reacting to the actions exerted on it. However, when these actions work negatively, water can lose its pleasant and welcoming characteristics. It becomes violent, destructive, and even mortal (Bachelard 1942). So water lets us live and, at the same time, it could give us death.

Currently, 7.9 billion people live on the Earth, and half of them live in urban areas, even if this percentage is expected to increase to 70% by 2050 (United Nations 2017). Although urban areas occupy only 3% of Earth surface (Ruth and Coelho 2007), they are contemporary the first responsible for and the first victims of climate change. Indeed, Man has become the most important agent of the global ecosystem (Harari 2015) because, pursuing progress, he has altered the fragile planet balances without considering the long times it needs to regenerate. For this reason, today, extreme meteorological events and urban floods are an inevitable fact that is increasing year by year, doubling compared to the eighties (Pernigotti 2015). In particular, the Centre for Research on Epidemiology Disasters (CRED) carried out a study on climate-related disasters showing how the latter dominate the temporal frame of the last twenty years. The CRED recorded a total of 7,348 events between 2000 and 2019. The study shows how 91% are closely related to climate change (CRED & UNISDR 2020). Among all, floods represent the most frequent disaster type (44%), affecting a total of more than 2 billion people and producing damages of 651 billion dollars during the same period (CRED & UNISDR 2020). For this reason, water has gained a negative connotation, so contemporary society has thought it would be better to stop, hide and send the water away as quickly as possible.

Starting from the abovementioned considerations and considering the needs of contemporary cities as well as the goals for the sustainable development (UN General Assembly 2015), this paper tries to investigate the role of the architectural and landscape project applying NBS to deal with the urban floods, providing a new urban resilience. In particular, the paper draws attention to the use of design strategies using the NBS and their benefits about the urban improvement from an environmental, social and cultural point of view, contrarily to the traditional approaches.

2   Nature-Based Solutions: Designing with Water

Shifting the Paradigm: From Resisting to Welcoming the Water

The European Directive 2007/60/EC on the assessment and management of flood risks highlights how human activities contribute to rise the flood occurrence possibility and, thus, worsen the consequences (Jha et al. 2012). The large number of floods and inundations that occurred in the world recently (CRED & UNISDR 2020) sparked the debate on the measures we should take to prevent human and material losses due to urban water invasions. Massive works, such as the Thames Barrier in London or the *Maeslantkering* in Rotterdam, represent the traditional approach based on resistance and defence from the water. But harnessing the water or building defence walls are not always the right strategy because, if on a hand they reduce the flooding probability, on the other hand they let the sense of protection to increase encouraging the buildings and human settlements close to the levees or dams. As a result, the potential damage increases as well as the exposure and the risk (Andreotti and Zampetti 2007). Nevertheless, there is who solicits new design strategies that have to go beyond the traditional idea to resist the water.

The "anti-barrier" advocates point out the need to shift the traditional paradigm (Marshal 2013) that needs to rely on the unpredictability of climate change so that urban environments could adapt and manage the flood risk more consciously. In general, they support urban design projects claiming a "new form" of resilience that should be both a response and action to maintain or find a new balance when the city must deal with a disorder or stress. In this perspective, Nature-based solutions represent a valid instrument (Mussinelli et al. 2018) because they offer exciting prospects in urban planning to deliver multiple benefits, reduce climate risks and improve dwellers wellbeing and urban life quality (Frantzeskaki et al. 2019). In particular, making space to nature (Merz et al. 2010) the blue-green infrastructures represent the opportunity of welcoming the water in the city and the possibility of designing with it, going beyond the mere mitigation but proposing both applied and theoretical reflection on how to obtain ecosystem services proactively (Mussinelli et al. 2018). Finally, the application of NBS in the architectural and landscape projects to deal with the urban floods implies "working with water" rather than controlling and "getting around the problem" (Grant 2012), and this gives the projects a pedagogical role.

The Architectural and Landscape Projects as a Medium for the Change
In this paragraph, four case studies are analysed briefly. Despite the differences in the geographical contexts and designers' culture, they all share a remarkable sensitivity towards the element of water and the principle of welcoming it and make it visible. Two of them have been completed (1 and 3), while two are still being implemented (2 and 4). Moreover, two of them refer to a broader climate adaptation and mitigation strategy (1 and 2), while the other two arise from different occasions, such as the needs dictated by the urban growth (3) and the old town refurbishment (4). They are clear examples of the possible coexistence between man and nature within an urban context. They show both structural and aesthetic-formal integration of NBS in the architectural and landscape project, highlighting the environmental, economic, social and cultural benefits they provide for a sustainable city future. A mutating and changing public space "welcoming" the risk and revealing it, improving stakeholders awareness, distinguish them. Finally, unlike traditional systems, they make public investments visible, building multisensorial spaces, integrating different functions and setting the basis for regenerating the physical and social fabric of the future city.

1-Waterplein Benthemplein by De Urbanisten, Rotterdam, 2013.

The first project here examined is the *Waterplein Benthemplein* that combines urban floods due to the cloudbursts with the urban voids improvement (De Urbanisten 2013). This project faces the flooding issue subverting the traditional approach to the problem. Here the flooding infrastructure is visible and not buried as the traditional ones. In the *Waterplein Benthemplein*, the violent water is not just a problem to solve but a pretext to improve the public space of an urban area and make the city adaptive and resilient to climate changes.

Fig. 1. (A) Waterplein Benthemplein, source http://www.urbanisten.nl - (B) The Soul of Nørrebro, source https://www.sla.dk/

The project consists of a square with three retention basins at different levels. When it rains, the basins welcome the water (Fig. 1A) thanks to a complex system of gutters. When the rain ends, the stored water goes towards a cleaning system before recharging the groundwater. When the square is dry, the *Waterplein Benthemplein* is a public space where people can do sport, play, relax and meet each other. The basins assume different functions such as skater court, theatrical or dancer stage, or even polyvalent sports field. Besides inside the square, there are a big fountain, a water wall and a water-well creating a pleasant water flow. However, during the heavy rains, this enjoyable flow becomes a rumbling waterfall that highlights visually and acoustically the water presence. Thus, the water becomes the essence of the square itself and the leading actor of an urban piece of art that is also an infrastructure (Klinenberg 2019). Finally, the project makes clear and enjoyable the public investment, and the water, from a problem, becomes a partner of the project that help to create new high-quality life spaces (Granata 2019).

2-The Soul of Nørrebro by SLA, Copenhagen, 2016.

The second project is the Soul of *Nørrebro*. It aims to convert water excesses from a problem due to floods and climate change into a resource to safeguard. The project consists of the renewal of *Hans Tavsens* Park (Fig. 1B), adding social, cultural and natural values to both the *Nørrebro* neighbourhood and the whole city. Using nature-based solutions, SLA converted the previous park into a "sponge-park", a sort of a wide retention basin capable of containing up to 18,000 m³ of rainwater that is cleansed by city nature biotopes along *Korsgarde* before it reaches the *Plebinge* lake.

The project mainly aims to answer the hydrogeological and climate change problems although, at the same time, it provides new urban spaces that improve sociality among the citizens and their relationship with nature (Cortesi 2017) and, in general, the quality of life. Thus, a strong synergy between the hydrological, biological and social cycles takes place in *Nørrebro*. Employing a humanistic and inclusive approach, the Soul of *Nørrebro* represents a design model for city development, based on the co-creation, dialogue and the design nature-based (SLA and Ramboll 2016). Thus residents receive a new public space and even the possibility of testing new ways of living with others, a new sense of belongingness, and a new sense of community (SLA and Ramboll 2016).

Finally, in *Nørrebro,* the water is part of the cityscape contributing to the public irrigation and improving the local microclimate through new blue-green infrastructures, as part of a climate change and flood resilience strategy.

3-Pop-Up Climate Change Adaptation Solution by Tredje Natur, globally, 2016.

The third sample is the Pop-Up Climate Change Adaptation Solution (Fig. 2A): a prototype thought in New York but scalable and globally adaptable. The Pop-Up represents a smart idea for adapting cities to the consequences of urban growth and, at the same time, climate changes. In particular, it has been designed mainly to handle the lack of parking facilities and green spaces and thus mitigating urban floods (Tredje Natur and Ramboll 2016). The Pop-Up is a project that changes the way of thinking to the architectural project radically. It is based on the concept of overcoming the separation of the infrastructural solutions to aim for a single project capable of maximizing the available surfaces, reducing investments and operating costs (Tredje Natur and Ramboll 2016). Hence, it is a unique solution to three of the most important challenges contemporary cities need to face.

Considering the Archimedes principle of flotation, the project consists of a structure that most of the time rests underground. It is only during heavy rains that it pops up and allows the water to be retained under it. Once the rainfall ends, the collected water flows out and is used for other urban uses. Thus, the Pop-Up solution appears in the cityscapes temporarily, only when cloudburst occurs, and its rising highlights the possible adaptation to the forces of nature. The round shape of the project makes the structure lighter, thus favouring buoyancy. Contemporarily, the parking spiral-shaped ramp lets to drive to and from the parking on the ground level, whatever the water quantity in the underneath reservoir. Pop-up represents a new way of thinking the urban architectural project where design and engineering try to deal with the needs for new sustainable urban ways of life and the climate change effects (Tredje Natur, Ramboll 2016). It is an attractive solution because it lets to save public money, urban space, and design more compact cities.

4-Banyoles Old Town refurbishment project by Miàs Arquitectes, Banyoles, 2008.

The last project proceeds from the rehabilitation of the canals system, the *recs,* that the monks of the monastery of *Sant Esteve,* the first human settlement in the city of *Banyoles,* build to avoid the floods coming from the nearby lake. The *recs* linked the lake *Banyoles* with the *Terri* river, crossing the historical downtown. Contemporarily to the floods management, they supplied water both for domestic and public uses. They provided water to the orchards and power the machines of the textile industry developed in this area. But, after the orchards and the factories were abandoned, over time, the network of canals gradually lost its original function and the overbuilding buried them below the town. So, a progressive urban degradation started, and the *recs* were merged into the city sewer system (Miàs 2008).

The proposal was designed with a double scope: enhancing the *Banyoles* historic centre and giving the *recs* their ancient function. The ancient system of historical squares that once struggled to open up in the old dense urban fabric now takes part in a new urban organization. They are connected by the streets where the existing sidewalks were demolished. Both squares and streets were stripped of the previous asphalt pavements and the historical town substrate has been shown, as well as the original network of the *recs*. The new travertine pavement is sculpted, depressed, even eroded, forming small basins or surface cuts where water can flow and interact with the people (Fig. 2B) (Miàs 2008).

The ancient *recs* now are part of the street and creates a new pedestrian, inclusive, and environmental and social high-quality public space. So, where once were cars and traffic jams, now children and adults stay, play and relax among the water streams that are both space qualifiers and means to control the urban runoff.

Fig. 2. (A) Pop-up climate change adaptation solution, source https://www.tredjenatur.dk/ - (B) Banyoles old town refurbishment project, source http://www.miasarquitectes.com

3   Conclusions

Nowadays, we do not lack technological tools to solve environmental problems; but even the most powerful technology might not be decisive if we ignore or reject the problem origin. Our cities have to face new challenges representing opportunities for creating and regenerating our cities more sustainably and smartly instead of a simple problem. From this point of view, under the architectural and landscape lens, this study examined urban flood risk and the use of NBS as an instrument to adapt and make more resilient the city, understanding nature, making space to water, welcoming and designing with it.

The paper tried to explain we should go beyond the common idea of water as an enemy destroying everything in its path. It points out how we cannot protect and defend our cities applying only "heavy engineering interventions" designing a demarcation line between what is dry and wet. Indeed, this generates a fake sense of security (Andretti and Zampetti 2007). Thus, it suggests we should apply soft blue-green solutions and water sensitive design strategies to reach that awareness that could let us achieve an anthropological, sociological and cultural mutation based on the sustainable symbiosis between Culture and Nature. Moreover, the case studies demonstrate how different approaches in architectural and landscape design using NBS can be perfectly integrated and coexist since they achieve similar objectives through diverse design processes. In the cases of Rotterdam and Copenhagen, the project stems from a request from above, dictated by a general adaptation strategy, conversely, in the other two cases projects arise from a not superordinate strategy. However, both approaches (top-down and bottom-up) in all four study cases intertwined. If a project having flood risk mitigation as the main goal uses a top-down approach, this becomes the starting point of a bottom-up approach application if we consider NBS benefits and the secondary objectives of the project itself. In fact, dwellers "receiving" new, improved urban places where they can meet, experiment with a new sense of belongingness and community, desiring and requiring more and better ones. Conversely, if a project arises from the bottom, it could become the occasion of meditating and planning the territory according to a new strategy concerning climate change and flood issues. So, integrating different approaches and objectives allows changing cities smartly, without wasting public money and land. In this way, by building multi-purpose and multi-objective solutions we will be able to make cities environmentally, economically, socially and culturally more sustainable.

Finally, it is time to debate on the need that our cities have to develop public spaces (Sepe 2021), and in particular, the green ones, improving or building them, even after the ongoing pandemic. Green areas support health objectives without degrading biodiversity, even better, giving the possibility to the biodiversity to spread out in a so "hostile" environment as cities could be. And, in this perspective, the architectural and landscape project can give us the reason to do it, making them more green or blue, liveable and sustainable from an environmental, social, economic and cultural point of view. Maybe we should remember and recover that ancestral bond with water and nature we had once because it might allow us to "reserve a place" where to live in the world of tomorrow.

References

Andreotti, S., Zampetti, G.: Le buone pratiche per gestire il territorio e ridurre il rischio idrogeologico. Legambiente (2007)

Bachelard, G.: L'Eau et les Rêves. Essai sur l'imagination de la matièr. José Corti, Paris (1942). Italian edition: Bachelard, G.: Psicanalisi delle acque. Purificazione, morte e rinascita (trans: Cohen Hemsi M, Peduzzi C). Rededizioni, Lavis (2006)

CRED & UNISDR: UNDRR Press release on the repot "The Human Cost of Disasters 2000–2019". United Nations (2020)

Cortesi, I.: Educare alla salute: gli spazi di relazione e movimento nella Natura Pubblica. Urbanistica Informazioni **272**, 910–915 (2017)

De Urbanisten: Plaza de Agua Benthemplein. Paisea **24**, 48–53 (2013)

Frantzeskaki, N., et al.: Nature-based solutions for urban climate change adaptation: linking science, policy, and practice communities for evidence-based decision-making. Bioscience **69**(6), 455–466 (2019). https://doi.org/10.1093/biosci/biz042

Granata, E.: Biodivercity. Città aperte, creative e sostenibili che cambino il mondo. Giunti Edizione, Firenze (2019)

Grant, G.: Ecosystem Services Come To Town: Greening Cities by Working with Nature. Wiley, Chichester (2012)

Harari, Y.N.: Homo Deus: A Brief History of Tomorrow. Harvill Secker, London (2015). Italian edition: Harari, Y.N.: Homo Deus. Breve Storia del Futuro (trans: Piani M). Bompiani, Milano (2017)

Jha, A.K., Bloch, R., Lamond, J.: Cities and Flooding. A Guide to Integrated Urban Flood Risk Management for the 21st Century. World Bank. (2012). https://openknowledge.worldbank.org/handle/10986/2241. Accessed 20 May 2021

Klinenberg, E.: Costruzioni per le persone. Come le infrastrutture sociali possono aiutare a combattere le disuguaglianze, la polarizzazione sociale e il declino del senso civico. Ledizioni, Milano (2019)

Marshall, A.: The \$5.9 billion question. Metropolis (2), 45–48, 80–82 (2013)

Merz, B., Hall, J., Disse, M., Schumann, A.: Fluvial flood risk management in a changing world. Nat. Hazards Earth Syst. Sci. **10**(3), 509–527 (2010). https://doi.org/10.5194/nhess-10-509-2010

Miàs, J.: Peatonalización y recuperación de las acequias. Casco antiguo de Banyoles. Girona. Paisea **7**, 98–103 (2008)

Mussinelli, E., Tartaglia, A., Bisogni, L., Malcevschi, S.: Il ruolo delle Nature-Based Solution nel progetto architettonico e urbano. Techne J. Technol. Archit. Environ. (15), 116–123 (2018). https://doi.org/10.13128/Techne-22112

Pernigotti, D.: Con l'acqua alla gola. Giunti Editore, Firenze (2015)

Roaf, S., Crichton, D., Nicol, F.: Adapting Buildings and Cities for Climate Change: A 21st Century Survival Guide. Elsevier, Oxford (2009)

Ruth, M., Coelho, D.: Understanding and managing the complexity of urban systems under climate change. Clim. Policy **7**(4), 317–336 (2007). https://doi.org/10.1080/14693062.2007.9685659

SLA: Ramboll: Climate Adaptation Copenhagen. The Soul of Nørrebro (2016). https://www.sla.dk/files/2914/9449/3217/SLA_Ramboll_HansTavsensPark_UK.pdf. Accessed 20 May 2021

Sepe, M.: Covid-19 pandemic and public spaces: improving quality and flexibility for healthier places. Urban Des. Int. **26**, 159–173 (2021). https://doi.org/10.1057/s41289-021-00153-x

Tredje Nature: Ramboll: Pop-up parking (2016). https://ramboll.com/~/media/files/rm/pop-up%20parking_english_long%20article.pdf. Accessed 20 May 2021

UN General Assembly: Transforming our world: the 2030 agenda for sustainable development. Report no. A/RES/70/1. United Nations (2015)

UN General Assembly: New urban agenda. Report no. A/71/L23. United Nations, Habitat III Secretariat (2017)

Reconciling Cities with Urban Nature: Towards the Integration of Ecosystem Disservices in Inclusive Spatial Planning

Hanna Skryhan[1][(✉)] and Anton Shkaruba[2]

[1] Institute of Environmental and Agricultural Biology (X-BIO), Tyumen State University, 25 Lenina Street, Tyumen 625003, Russia
a.y.skrigan@utmn.ru
[2] Institute of Agricultural and Environmental Sciences, Estonian University of Life Sciences, Kreutzwaldi 5, 51006 Tartu, Estonia
anton.shkaruba@emu.ee

1 Introduction

Functions and properties of ecosystems delivering discomfort to citizens, also known as *ecosystem disservices* (EDS) (Döhren and Haase 2015; Lyytimäki 2014) are fundamentally important in terms of interactions between people and urban nature (Dobbs et al. 2014; Escobedo et al. 2011; Lyytimäki 2014; Vaz et al. 2017) and can be at least as important for citizens as ecosystem services (ES). This further leads to the call for the solutions whereas ES as well as EDS are integrated in planning designs delivering comfortable urban environment to citizens (Blanco et al. 2019; Vaz et al. 2017). This however represents a major challenge not only due to many trade-offs, such as choices between space and commercial development benefits vs. ES (Spyra et al. 2020), but also due to conflicting perspectives and preferences of various stakeholder, epistemic and social groups on the very nature of EDS and ES (Shkaruba et al. 2021). We assume that there are at least two compelling reasons for EDS to be addressed by the planning process in its broad sense. This is for urban nature in order to survive, and for citizens in order to benefit from the services it provides. This needs to entail the formulation of multistakeholder consensus over EDS/ES, and ideally to consider the broadest possible variety of interest groups (including age and gender) and possible conflicting perspectives. Working to address this challenge, this research focuses on EDS in urban communities, and explores them in terms of inclusive planning.

Inclusive planning stems from the policy concept of 'social inclusiveness' developed in order to recognise and consider in policies and management practices the diverse needs and abilities of people (Meyer and Hinchman 2007). Due to diverse perspectives on urban nature, this concept received attention of urban planners and managers seeking to ensure accessibility and social acceptance of ecosystems incorporated to the fabric of sprawling cities (Van Herzele et al. 2005; Roth et al. 2017), and lately also ES and NBS (Nature 4 Cities 2020; Schaubroeck 2017; Van Herzele et al. 2005). Based on a comprehensive stakeholder analysis carried out over the past decade in the city of Mahilioŭ (Belarus) we bring forward a decision making tool that can help to identify EDS and to choose

© The Author(s), under exclusive license to Springer Nature Switzerland AG 2022
D. La Rosa and R. Privitera (Eds.): INPUT 2021, LNCE 242, pp. 64–72, 2022.
https://doi.org/10.1007/978-3-030-96985-1_8

appropriate strategy for the development of green and blue infrastructure (GBI) that would address them in an inclusive manner. The tool is explained using three cases of representing different EDS-related contexts.

2   Methods and the Case Study Area

The decision tree has been constructed inductively based on 32 interviews with various stakeholders and 6 stakeholder workshops held between 2010 and 2021 over the city's GBI and its ES and EDS-related planning issues in the valley of the river of Dubravenka in Mahilioŭ (Fig. 1). Mahilioŭ is the 3rd largest city on Belarus, with population over 380,000, administrative area round 120 km^2, with a historical core edged by rivers of Dnieper (Dniapro) and its small tributaries (including Dubravenka), and large industrial areas on city's outskirts.

As a starting point, we have developed an EDS classification accounting for any EDS found in temperate climates of Eastern Europe; the classification is summarized in the Table 1. It is based on 3 rounds of sociological polls hold in the city in 2016–18 (total over 900 respondents) addressing various aspects of GBI management, and on literature and authors' observations covering cities in the region (Belarus, Russia, Ukraine). Its first draft went through several discussion rounds with stakeholder representatives (such as professional planning and management communities) to ensure its comprehensive character, relevance, clarity and logical structure. As the next step, the tree was drawn following the analysis of EDS management cases observed by authors in Mahilioŭ and other cities in the region. Likewise, the tree was revised following discussion rounds with stakeholders. The agreed version is set in Fig. 2; its application is explained using 3 distinctive "focal action situations" (Ostrom 2009) (or simply "situations" from now on).

3   Deployment of the Decision-Making Tree

The river of Dubrabenka and its valley stretch for over 10 km across most of Mahilioŭ with the valley reaching 600 m wide. Historically this landscape had brought massive benefits to citizens. With the city growing and production and lifestyle systems changing, the river fell in disregard, and became considered as the city's dirty backyard. Only the recreational function has survived in some areas. The recent decade is marked with attempts to make use of this landscape. So far most of the attempts had to do with destroying the ecosystem function and character in order to make the area suitable for high story developments, or to convert it to walking areas dominated by sealed surfaces. Yet Dubravenka can offer much more, but for this the citizens would need to rethink quite a bit their zone of comfort. This process needs to be adequately helped by information, communication, management and technology. In what follows we will review 3 situations illustrating how the decision tree helps or would help to find solutions.

Table 1. A classification of ecosystem disservices

EDS group	EDS sub-group	EDS examples
I. Ecosystem attributes and functions	Ia. Ecosystem attributes	"Unacceptable" ecosystems (for example wetlands), invasive species
	Ib. Events generated by urban ecosystems	Floods, landslides, erosion, forest, grassland or pit bog fires
	Ic. Functioning of urban ecosystems	Harm from bird excrement on artificial surface, risks of falling old trees and branches, leaf litter, seeds and pollen and etc.
II. Human health	IIa. Risks related to human health	Allergies and diseases, hygiene and health problems, toxic species, biting animals and attacks by wild animals
	IIb. Nature related fears	Fear of wild animals, fear of darkness, fear of wild nature in general
III. Aesthetic issues	N/A	Loud voices of birds, dogs, and etc., excrement in green areas, unmanaged green areas, presence of gulls, mosquitoes, mugwort or nettle, unpleasant smell
IV. Restrictions and inhibition of urban planning and development	IVa. Restrictions caused by nature protection	Protected species and areas inhibit planning and construction
	IVb. Inhibition of activities	Crimes connected with urban parks, poor condition of unpaved pads, shade and visual obstacles from vegetation, block of transport connectivity

Situation 1: what opportunities for mindset transformation are worth considering, but were not so far

In its middle 3 km long stretch Dubravenka is adjoined by a 50–300 m wide floodplain full of wetlands, cut-offs and floodplain ponds. This ecosystem demonstrates vivid natural dynamics, and it is rich in animal and plant diversity. The floodplain is closely adjoined by multistorey blocks and by areas of wooden cottages with gardens. Till the late 1990 Dubravenka's meadows were still used to harvest fodder and to graze the

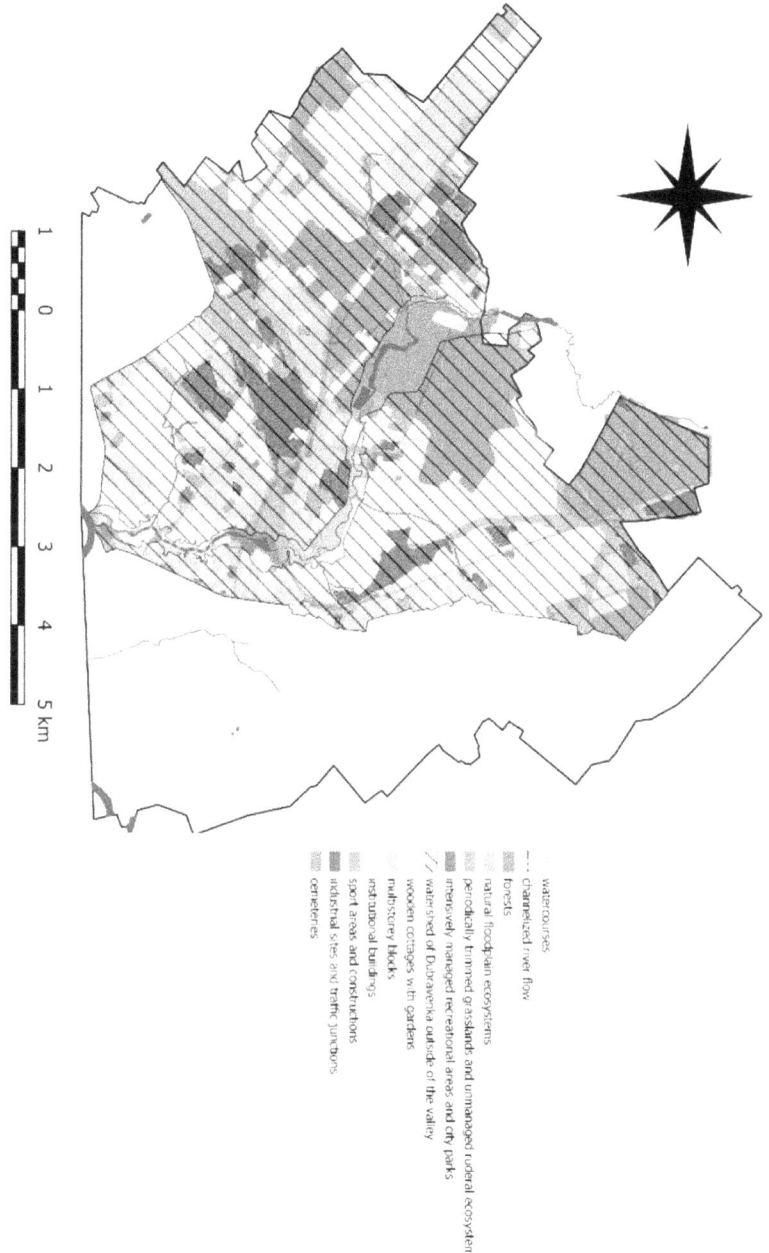

Fig. 1. The watershed and the valley of Dubravenka within the administrative border of Mahilioŭ

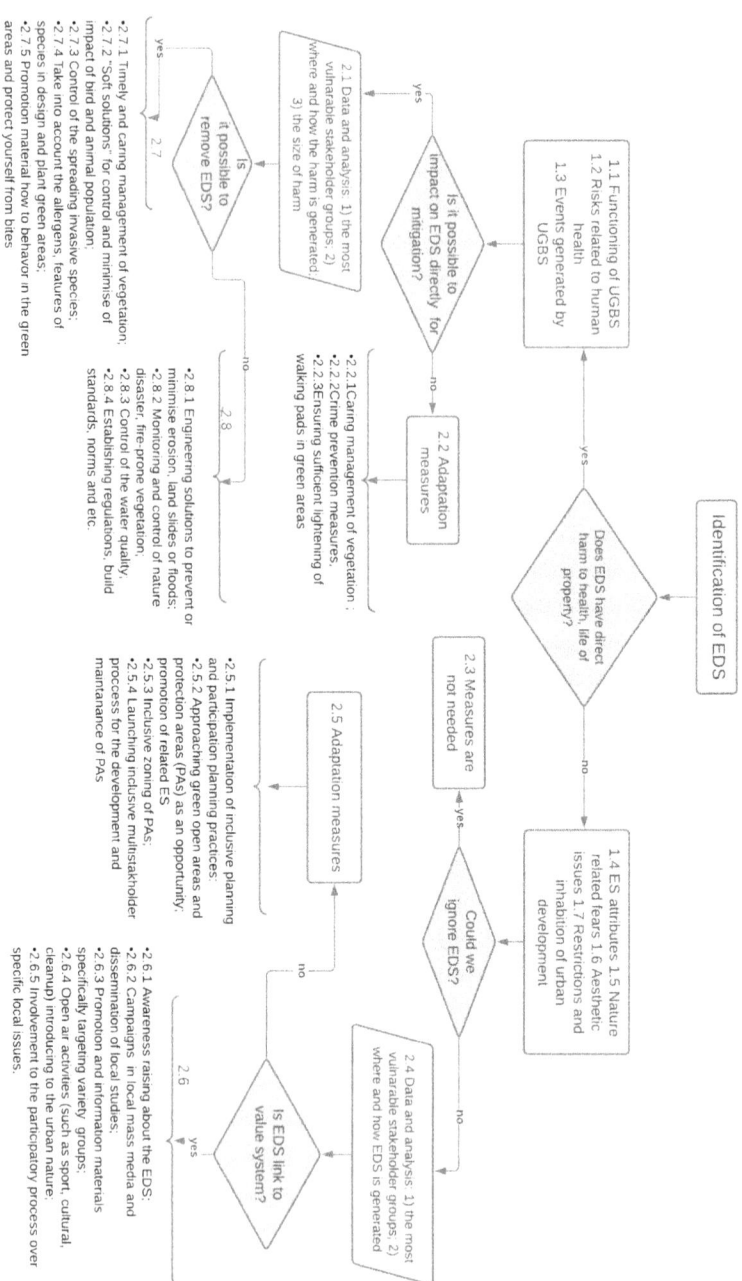

Fig. 2. The decision-making tree for the identification and management of EDS in cities

goats. This practice has been gradually abandoned, and once it was gone, any advantages of living by the river had disappeared. Most of the citizens living nearby complain of insects, smell, fogs, dirt, access and connectivity issues, overall feel of marginalization, and aesthetically upsetting scenery. The rest of the city shares the general sentiment of the neighborhood. This attitude is also revealed and illustrated through the common discourse about a stretch of the floodplain that had been destroyed in the late 1980s. The original ecosystem was a mangrove-like periodically flooded alder grove, with the river divided there in many streams running on a sand bed. By the 2000s it became a system of ponds with embarkments, fountains and open spaces with mowed lawns and sealed surfaces. While this action had created access to the river in a manner appreciated by citizens, it is also noteworthy that its original character or any positive associations were not recorded in the city's memory.

The city needs to develop, and the options currently looming ahead are re-development of the floodplain in yet another lifeless open space. The options providing for the preservation of biodiversity and ecosystem character are either (1) keeping *status quo* in its current or perhaps marginally improved state (e.g. providing better access to the sides of the floodplain and ensuring safety and better maintenance of the neighborhood), (2) providing any possible access to and through the ecosystem (e.g. by constructing walking pads, observation towers (e.g. for bird or beaver watching), preparing information and creating learning infrastructure) in order to articulate the learning and create new recreational value, or (3) establishing a conservation regime with wildlife protection prioritised, and non-destructive observation and learning opportunities arranged. Any of these options conflict with current planning policies, priorities, practices or citizens' expectations, although they are the most rational ones in terms of the making the most of the ecosystem's value (including the resilience to climate change), and would bring a new and potentially appreciated dimension to the city's GBI. The option (1) might be more suitable for the current situation of resource scarcity, while the (2) should be possible only if high quality expertise and considerable development costs are secured.

Assuming that the municipality would take a planning decision that would be strategical in terms of the development of a livable urban environment, the decision-making tree would suggest the following action for ensuring the inclusive planning process. The EDS described in this situation belong to groups 1.4.1, 1.6.3, 1.6.4 and 1.6.5 (Fig. 2). Within the paradigm of inclusive planning, the most exposed group would be the citizens living next to the floodplains (2.4), because floodplain-related EDS do not spill over this area. The next steps entail planning of communication strategies for the modification of value systems (2.6) and management actions to develop the options enhancing the acceptance of new developments (2.5).

Situation 2: how citizens actively resist when communication was not convincing

In 2015–17 an environmental NGO was implementing a water purification project including a small oil trap and a constructed wetland in a mouth of a ravine with a discharge point of stormwater sewers. The people living in nearby cottages had been alarmed by a news of a "water-treatment facility" being constructed in their backyard and opposed to it, although the NGO activists did their utmost to convince them that the wetland is very small scale, would not add anything to the existing emission, and would be designed as a nice landscaping feature. The activists had undertaken a massive cleanup of ravine that was full of garbage, however that only made the neighbors even more suspicious. Being terrified by the "water-treatment facility", including its EDS component, they had directly appealed to the city mayor, who ordered to halt the project, although such an intervention was in breach of any regulations and procedures.

This situation is interesting as a clear scenario for a top-down action aiming to introduce GBI featuring EDS: any such GBI would have perfect chances to be overruled through an appeal to higher decision-making floor, because the expected political gain would always outweigh the perceived value of such a project. If consulted prior to the project, the decision-making three (Fig. 2) would suggest that the EDS described here had to do with the aesthetic issues (1.6.4 and 1.6.5). The same as in the previous situation, the most exposed group are residents (2.6) worried about possible smells, insects, upsetting scenery and air pollution. On the next step developers had to check options from the boxes 2.6.1 and 2.6.5.

Situation 3: what adaptation of mindsets took place, and what hopes it gives

Piačersk woods is a historical city forest of Mahilioŭ with its core in Dubravenka valley. With an eponymous large pond in its southern zone, it is an important multifunctional recreational zone. This generates considerable footprint on the ecosystem that has lost much of its natural character in a large part. However, it also managed to preserve an impressive mosaic of vegetation types, as well as several red-listed species. Since early 2000s a group of local NGO activists and academics had been working on making a case for establishing a protected area in Piačersk woods. Since the very beginning this represented a monumental challenge. As the first stakeholder workshop (2011) demonstrated, the citizens (even potentially informed ones) did not recognise conservation value of the forest. Gradually a multistakeholder community appreciating and exploring the value of Piačersk has evolved, including concerned residents, owners of small business, academics, journalists, planning and environmental professionals, as well as broader NGO circles. In 2018 the activists solicited a study visit by experts from the National Academy of Sciences, who confirmed the conservation value of Piačersk and formally notified the local authorities that a protected area needs to be established there. This information was very well received by citizens, who appeared to be ready to accepts conservation restrictions in exchange for better protection. Although the municipality was somewhat reluctant, it could not see opportunities to back up their reluctance by popular opinions, and therefore it has acknowledged the requests and started necessary proceedings.

In this situation all the conditions for turning a potential EDS to valuable ES have been met. The EDS belonged to 1.5.3, 1.7.1 and 1.7.2, and the groups with the greatest potential objectives would be the citizens residing around the forest or visiting it, as well as the municipality and affiliated developers. Turning disadvantages to benefits took value systems and core beliefs to modify, and in order to achieve this the activists started with options 2.6.1 and 2.6.2, including excursions, publications on the Piačersk's value for the current and future generations, and promoting the forest to academics as a research ground. In parallel, the group of activists explored 2.6.3 and organized 6 stakeholder events to share the mission with more parties. As a result, more spin-off initiatives emerged (2.6.4) organized by business (e.g., Piačersk-themed gastro fests and folk fairs), sport clubs (e.g., canoeing, diving or sport-orienteering events with a litter-collection component) and other environmental activists organising regular clean-up or tree planting fests). The latter worked particularly well to improve understanding of the municipality and the locals. Pursuing 2.6.5 the activists tried to institutionalise the questions and to push it to the local policy agenda through meetings with decision-makers and petitioning to higher policy floors. Finally, a dedicated committee with some NGO representatives (unusually for Belarus) included has been set up, and the organization of the protected area became a practical matter.

4   Discussion and Concluding Remarks

The experiences of developing and testing the decision-making tool (including the EDS classification) in Mahilioŭ have demonstrated that promotion of urban nature is, in most cases, is a problematic affair, and to achieve stakeholder acceptance it requires timely deliberation based on the understanding of stakeholder perspectives, and at a pace that is acceptable to them. Our decision making tree can be used as a tool supporting the inclusive planning process or, just ensuring the level of acceptance that would let the project move forward.

The Belarusian context poses some specific challenges that might be typical for other societies in transition and that are getting aggravated in societies with top-down management cultures. In examples from China (Wang and Wang 2020) we have recognized the issues typical for Mahilioŭ, such as: (1) the emergence of unexpected stakeholders that obscure inclusiveness and transparency principle of communicative planning; (2) the hidden connections that deliberately exclude the planning and design team; and (3) value conflicts that distorted the effectiveness of communicative actions. Our evidence suggests that a sensible deliberation may help to overcome these hurdles, and that the role of social learning cannot be overestimated, also because it is, at the end of the day, cheaper than most other options. As also underlined by Apfelbeck et al. (2020), mutual trust building with local communities is the best investment, and once this is achieved, the skepticism of local elites is easier to break, so the planning action can be duly institutionalised and pushed forward. Fundamentally, the country contexts like Belarus (and possibly also China etc.) need some space (still lacking) where planners could interact with various stakeholder circles, where all the voices are taken in a constructive manner.

As we have demonstrated, and similarly to the conclusions of Blanco et al. (2019), the concept of EDS is fully operational as a planning instrument. EDS indeed need to be considered in their specific socio-cultural context, accounted for their biophysical nature and time scales, for the implications for space allocation, as well as for the trade-offs with and "conversion potential" to ES. These considerations have been incorporated to the EDS classification and the decision-making tool, and further illustrated by the three situations. The most challenging step is understanding why and by whom exactly (including the root causes) urban nature is perceived as EDS. Once the planning process is ready for inclusive approaches, the rest should come relatively easy.

Acknowledgments. This work was supported by the EU Erasmus+Jean Monnet Module EARTH (No. 600452).

References

Apfelbeck, B., et al.: Designing wildlife-inclusive cities that support human-animal co-existence. Landsc. Urban Plan. **200**, 103817 (2020). https://doi.org/10.1016/j.landurbplan.2020.103817

Blanco, J., Dendoncker, N., Barnaud, C., Sirami, C.: Ecosystem disservices matter: towards their systematic integration within ecosystem service research and policy. Ecosyst. Serv. **36**, 100913 (2019). https://doi.org/10.1016/j.ecoser.2019.100913

Dobbs, C., Kendal, D., Nitschke, C.R.: Multiple ecosystem services and disservices of the urban forest establishing their connections with landscape structure and sociodemographics. Ecol. Ind. **43**, 44–55 (2014). https://doi.org/10.1016/j.ecolind.2014.02.007

von Döhren, P., Haase, D.: Ecosystem disservices research: a review of the state of the art with a focus on cities. Ecol. Ind. **52**, 490–497 (2015). https://doi.org/10.1016/j.ecolind.2014.12.027

Escobedo, F.J., Kroeger, T., Wagner, J.E.: Urban forests and pollution mitigation: analysing ecosystem services and disservices. Environ. Pollut. **159**, 2078–2087 (2011)

Haase, D., et al.: Greening cities – to be socially inclusive? About the alleged paradox of society and ecology in cities. Habitat Int. **64**, 41–48 (2017)

Lyytimäki, L.: Bad nature: newspaper representations of ecosystem disservices. Urban For. Urban Greening **13**(3), 418–424 (2014). https://doi.org/10.1016/j.ufug.2014.04.005

Meyer, T., Hinchman, L.: The Theory of Social Democracy, p. 288. Polity, Cambridge (2007)

Jeuken, Y.R.H., Breukers, S., Sari, R., Rugani, B.: Nature 4 Cities: Nature Based Solutions: Projects Implementation Handbook (2020). https://55d29d92-2db4-4dd3-86b5-70250a093698.usrfiles.com/ugd/55d29d_58487570de4147d38035351ba5b61a69.pdf

Ostrom, E.: A general framework for analysing sustainability of social-ecological systems. Science **325**(5939), 419–422 (2009). https://doi.org/10.1126/science.1172133

Roth, D., Vink, M., Warner, J., Winnubst, M.: Watered-down politics? Inclusive water governance in the Netherlands. Ocean Coast. Manag. **150**, 51–61 (2017). https://doi.org/10.1016/j.ocecoaman.2017.02.020

Schaubroeck, T.: A need for equal consideration of ecosystem disservices and services when valuing nature; countering arguments against disservices. Ecosyst. Serv. **26**, 95–97 (2017). https://doi.org/10.1016/J.ECOSER.2017.06.009

Shkaruba, A., et al.: Development of sustainable urban drainage systems in Eastern Europe: an analytical overview of the constraints and enabling conditions. J. Environ. Plann. Manag. **64**, 2435–2458 (2021). https://doi.org/10.1080/09640568.2021.1874893

Spyra, M., La Rosa, D., Zasada, I., Sylla, M., Shkaruba, A.: Governance of ecosystem services trade-offs in peri-urban landscapes. Land Use Policy **95**, 104617 (2020)

Vaz, A.S., et al.: Integrating ecosystem services and disservices: insights from plant invasions. Ecosyst. Serv. **23**, 94–107 (2017). https://doi.org/10.1016/j.ecoser.2016.11.017

Van Herzele, A., De Clercq, E.M., Wiedemann, T.: Strategic planning for new forests in the urban periphery: through the lens of social inclusiveness. Urban Forest. Urban Greening **3**, 177–188 (2005)

Wang, C., Wang, Y.: In search of open and inclusive arenas: transnational practice of communicative planning in Yongtai, China. Habitat Int. **106**, 102288 (2020). https://doi.org/10.1016/j.habitatint.2020.102288

Designing Cultural Ecosystem Services in Territorial Planning. The Case Study of Ancient *Burgentia* (Basilicata, Italy)

Francesca Perrone[1]([✉]), Maria Teresa Palma[2], and Duilio Iamonico[1]

[1] Department of Planning, Design and Technology of Architecture (PDTA), Sapienza University of Rome, Rome, Italy
francesca.perrone@uniroma1.it
[2] Freelance Archaeologist MIC, Specialist in Late Ancient and Medieval Archeology, SSBA of Matera, Matera, Italy

Abstract. The paper, as the result of an interdisciplinary research study, analyzes how the reconstruction and the practical feasibility of a historical-environmental Cultural Ecosystem Service (CES) can facilitate and implement the functional development of territorial and landscape planning tools. CESs depend on the stability of landscapes, for both the "direct" protection and enhancement of the identity heritage and the "indirect" preservation of the historical memory. Yet, plans barely focus their attention on this subject, from both conceptual and methodological points of views. In order to highlight the importance of an interdisciplinary cognitive-methodological path, which sheds the light on the complexity of the historical-cultural and ecological-environmental conditions of a site, we carried out an analysis of the ancient *Burgentia* (Basilicata region, Southern Italy), a historical-cultural and environmental site, whose area is adjacent to the territory of the Appennino Lucano Val d'Agri Lagonegrese National Park. The ecosystem functions, which allowed to recreate the identity of the place, underlining the features that characterize it as a CES, belong to the following three components: physical-settlement-geographical-geomorphological, botanical, and historical-archaeological. The main aim of the study is to propose a method for identifying and enhancing CESs to be included in the set of planning tools. Plans should allow an in-depth knowledge of the ecosystem-territorial components, and above all they should improve their use and management in a sustainable way, fostering the transition from a recognized cultural heritage to a fully accessible CES. At the same time, the focus on these Ecosystem Services can hold the position of plan engine for every sector of intervention, producing a virtuous two-way relationship. The studies carried out let us integrate the elaboration of specific in-depth thematic analysis of the fact-finding survey and support the governance model established by the Park Plan. The goal is to boost both the Park Plan function of collecting data and its active role for the development of cultural heritage, infrastructural networks and context services, which are necessary for the development of the peculiarities of the places.

© The Author(s), under exclusive license to Springer Nature Switzerland AG 2022
D. La Rosa and R. Privitera (Eds.): INPUT 2021, LNCE 242, pp. 73–80, 2022.
https://doi.org/10.1007/978-3-030-96985-1_9

1 Introduction

Ecosystem services (ESs) are "the conditions and processes through which natural ecosystems, and the species that make them up, sustain and fulfill human life" (Daily 1997, p. 3). Although difficult to quantify, the identification of the ecosystem services are increasing, since they are necessary for the characterization of landscape identity, environmental sustainability, development, etc. The quality of these services and their functions support the quality of life of living beings. The increased awareness of man's dependence on natural factors (and vice-versa) means that researchers and planners are increasingly interested in the conceptual challenges on the ES, to transfer into operational tools (strategies and measures) that are suitable to the sustainable development of the landscape (Plieninger et al. 2013). ESs are distributed into four distinct categories: supporting services, regulating services, provisioning services and cultural services. Cultural Ecosystem Services (CESs), like "tangible and intangible goods and chattel" (Vasiljevic and Gavrilovic 2019) can, in turn, be further union/combination of:

- Services for the "direct" defence of identity heritage and socio-cultural values (geological, paleontological, archaeological, etc.);
- Services for the "indirect" protection of the genius loci, historical memory, the aesthetics of the landscape and the spirituality of the places.

The CESs received, unforunately, little attention, mainly due to a inability to distinguish their types and to evaluate its socio-cultural importance. This fact depends on how intangible or hardly tangible (for the biophysical system) and quantifiable (for monetary value) ecosystem services are. Their identification and surveying are lacking, as information deriving from several disciplinary fields (history, archeology, pedoarchaeology, ecology, social sciences, economics, etc.) is difficult to integrate. Furthermore, the "tacit" values of CESs (Milcu et al. 2013), even when identified, strongly depend on other ESs and on the possibility to benefit from them.[1]

Given that "there are still conceptual and methodological gaps in the realm of CES" (Maraja et al. 2016, p. 33), difficulties persist in the political-decision-making processes that should derive from that (Albert et al. 2016).

The aim of this study is to increase the knowledge of CESs towards specific and detailed ways of identifying, classifying, and quantifying their goods and benefits derived which can be included in the set of plan tools (Plieninger et al. 2015). The plan is, in turn, necessary to allow a virtuous use of the services and, at the same time, to give to ES the role of the plan engine for all other sectors of intervention. Since plans involve services and the latter structure the plan, it is necessary, "in the time of homogenization and disappearance of the specific character of the landscape" (Vasiljevic and Gavrilovic 2019) before approving the socio-cultural enhancement and tourist-recreational use of an area, to pursue objectives of awareness of the opportunities linked to territorial resources.

---

[1] Stålhammar and Pedersen (2017) adopted an "interpretivist approach" to deepen the study of CES based on the perception experienced by the users, defining three conceptual implications (described by the participants in the focus group interviews): 'indivisibility', 'incommensurability', and 'the goodness of perceived naturalness'.

The conceptual reconstruction of a cultural benefit (CES) serves as an important step for setting up adequate and efficient planning processes, which focus on increasingly inclusive strategies (Plieninger et al. 2015). Together with the studies of the experts, the acquisition of a greater awareness of CES can be useful for bringing together the different conservation programs of the naturalistic and cultural heritage (consisting of cultural and landscape heritage, art. co.1, Code of cultural heritage and landscape 42/2004), improving the results and increasing the political-decision-making support (Plieninger et al. 2013).

To achieve the above mentioned aim, we selected the area of the ancient medieval site named *Burgentia* [located into the Brienza Municipality (Basilicata region, Southern Italy), and adjacent to the Appennino Lucano Val d'Agri Lagonegrese National Park] which is not still adequately evaluated despite its potential in providing ESs [see also the architectural panels of the Park Plan titled "Paesaggi: unità e componenti generative" (= Landscapes: units and generative components) and "Strategie di interpretazione e valorizzazione delle Risorse naturalistiche e culturali" (= Strategies for interpretation and exploitation of the natural and cultural resources].

2 Material and Methods

The work was carried out using an interdisciplinary approach reflecting the complexity of the landscape of the study area which includes historic, cultural, and environmental aspects.

The identity of the study area was defined by the identification of their CES which was, in turn, classified into the following three components:

Physical-settlement, geographical, and geomorphological component
We analyzed the specific literature (see below in the paragraph results) concerning both the geological, geomorphological, and idrographic aspects. The chemical-physical analyzes of the soil (pH, electrical conductivity, salinity and total limestone (Giandon and Bortolami 1990) were carried out according to the methods reported in the Ministerial Decree Official methods of chemical analysis of the soil 13/09/1999[2].

Botanical component
Field investigations were carried out during the spring of the year 2018. The collected and dried plant specimens are stored in the Herbarium Flaminio (international code HFLA, in accordance with the Index Herbariorum of New York) (Thiers 2021 [continuously updated]). The taxa are identified using the new Edition of the Flora d'Italia (Pignatti 2018). The nomenclature follow Bartolucci et al. (2018) for the native species and Galasso et al. (2018) for the aliens.

Historical-archaeological component
On the basis of our careful investigations, no archaeological study regarding the studied area was traced. The data completely derived from field surveys.

---

[2] We thank the Department of Crop Systems, Forestry and Environmental Sciences. University of Basilicata (Faculty of Agriculture) and Professor Laura Scrano, our tutor for the research, for processing the analytical data.

3   Results

Physical-settlement, geographical, and geomorphological component

The area analyzed within the study is located near the fortified site, consisting of the Caracciolo Castle and the adjacent medieval village, located on a hill of difficult access, overlooking the valleys below, where today's town of Brienza develops (Fig. 1). It is adjacent to the Appennino Lucano Val d'Agri Lagonegrese National Park, representing a good case of historical-cultural and landscape bridge with the nearby Campania region (to north).

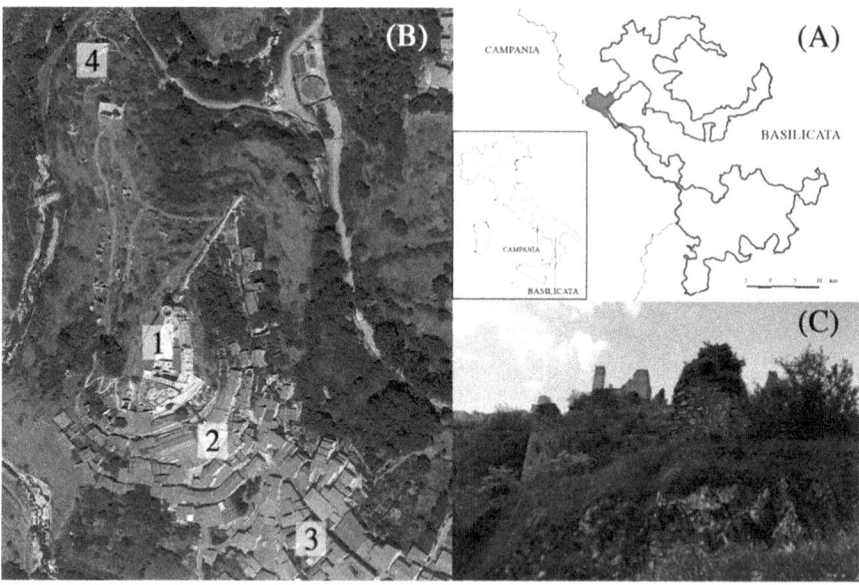

Fig. 1. Brienza town and adjacent areas. (A) Brienza Municipality (in red) within the Appennino Lucano Val D'Agri Lagonegrese Park (green line); (B) satellite image of the medieval site within the Municipality of Brienza (Potenza andimistrative Province): 1) Caracciolo Castle, 2) Medieval village, 3) current residential area, 4) study area; (C) picture of a general view of the study area.

The territory of Brienza (the ancient *Burgentia*) is crossed by two streams, named Fiumicello and Pergola, which, right next to the study area, join together, then flow into Melandro and Tanagro rivers, tributaries of Sele river. From the geologica point of view, the area is located along a tectonic line with a NW-SE trend, highlighting an overlap of the lands of the Apennine Platform on those of the Lagonegrese Basin. The stratigraphic-structural units that emerge within the site belong to the stages of the Cenozoic tectonics and from a mineralogical perspective they belong to the calcareous/siliceous/marl series. The site is characterized by mountainous limestone soils "with predominantly lithology identification in carbonate formation (limestone, dolomitic limestone, oolithic lime-stone), and subordinately turbidite (siliceous marl, clayey marl and sometimes quartz-micaceous sandstone). Natural plant formations prevail (grasslands, sparse woods), used for grazing and passing underneath high-trunk forests of deciduous and locally coniferous trees" (Doddato 2017, p. 24).

Concerning the soil is characterized in having a moderate alkalinity (pH between 8.33 and 8.47) which reflects the so-called "constitutional alkalinity" typical of calcareous soils. Data is confirmed by the total limestone, equal to about 56% for one sampling point and 81% for the other sampling point, of the site under consideration. The electrical conductivity (EC) is between 0.57 and 0.61 µς/cm, identifying a soil whose EC level does not show any particular danger of crop depression (Giandon and Bortolami 1990, p. 15).

Botanical component
The study area is physiognomically and floristically characterized by mesophile grassland dominated by members obelonging to the families Poaceae [e.g., *Avena barbata* Pott ex Link, *Bromopsis erecta* (Huds.) Fourr. subsp. *erecta*, *Dactylis glomerata* L. subsp. *glomerata*, *Phleum pratense* L. subsp. *pratense*, *Poa bulbosa* L. subsp. *bulbosa*, *Poa pratensis* L. subsp. *pratensis*)] and Fabaceae [e.g., *Medicago arabica* (L.) Huds., *M. polymorpha* L., *M. sativa* L., *Trifolium spp.*] along with mainly *Anthemis cretica* L. subsp. *cretica* (Asteraceae), *Daucus carota* L. subsp. *carota* (Apiaccae), *Papaver rhoeas* L. (Papaveraceae), and *Salvia pratensis* L. (Lamiaceae). There are also scattered indivuals of shrubs (*Cornus sanguinea* L., *Rosa canina* L., *Spartium junceum* L.) and trees (*Euonymus europaeus* L., *Fraxinus ornus* L., *Quercus ilex* L. subsp. *ilex*).

Historical-archaeological component
The site is characterized by buildings which prove the presence of humans during historic age, as well as highlighting its socio-cultural identity. In particular, there is a medieval fortify calstle (*castrum*) at the top of the site, whereas the ancient village occur along the slopes. The village is located following the natural morphologies of the substrate which is characterized by high topographics differences in altitudes. This urban configuration is clearly related to the need to safeguard the population from attacks. Furthermore, the location of the village and clastle is strategic to manage the sociopolitical and economical communications (by the mobility) of the Normandy-Suevian age.

4   Discussion

As a whole, both the biotic and anthropic features of the study area highlights the constant presence of humans in the course of the time. For example, the occurrence of mainly mesophile grasslands as well as the richness of Poaceae and Fabaceae species (floristical-physionomic characterization) suggests that the anthropic actions caused a change the landscape structure, i.e. favoring the spreading the natural mesophile grassland in place of pastures. In fact, grazings represented one of the main economic activity in the past in both the Brienza territory [as stated in the unpublished fourteenth-century document (AC, Arm. X, Ms 6) found by us in the "*Inventarium bonorum S. Jacobi et S. Johannis de Burgentia*"] and the whole Basilicata region (Pedio 1964, 1965). To note that this landscape changing occurs also in other Italian areas which are structurally similar to the *Burgentia* site, e.g. the archeological-naturalistic Appia Antica Regional Park in

Rome (Iamonico 2008) where the partial abandonment of the agricoltural activities (i.e. grazing and cultivations) caused an increasing of natural herbaceous comunities. This fact, is surely good if from the ecological point if view, isnce the environmental quality of the landscape increase (see e.g., Iamonico and Di Pietro 2018). However, the economic aspect (i.e. that related to the agricoltural activity) drastically decreased. Furthermore, it is to be noted that the ruins are almost completely covered with authoctonous liane *Hedera helix* L. subsp. *helix*, which is potentially harmful to the artifacts it tends to penetrate, causing, through the diametrical growth of the trunks, enlargement and possible breakage.

In this view, we can state that both the natural and the archeological resources are not adequately used as ES or CES in the study area, whereas that this use could be excellent to recover the economic activity of the whole territory of Brienza Municipality. It could also act as a catalyst for the analysis of crucial sectors of the Park's usability that could no longer be postponed, further increasing knowledge and overall awareness.

5   Conclusion

The Plan of the Appenino Lucano-Val D'Agri-Lagonegrese National Park is under approval and could benefit from a programmatic-planning change aimed at enhancing and using the CESs.

The Plan for the Park could play a significant role of "dynamic informer" (as for the elaboration of the documents) for those areas of considerable historical and environmental interest. In a broad and systematic project for the protection and enhancement of resources, it is necessary to document and raise awareness of the cultural heritage and, consequently, the cultural services it provides. In order to increase the knowledge of protected areas, as well as to improve their management, it is necessary to document the various aspects of the landascape, i.e. biotic and anthropic aspects and, when necessary, to implement the rules of the Plan.

The framework law on protected areas 394/1991 provides that the Park Authority, through the plan, must regulate "the protection of natural and environmental values and, in particular, the general organization of the territory and its classification in areas or parts characterized by different forms of use, enjoyment and protection" (art. 12, par. 1, lett. a) (see also e.g., Di Pietro et al. 2019). That is imagined in a perspective of sustainable and integrated development of ecological-environmental and historical-cultural components belonging to a single territory.

The here proposed in-depth analysis can help to increase knowledge of the characteristics of the area and, consequently, to define the degree of protection it requires and the allowed activities of enhancement, use and development.

Ultimately it is possible to hypothesize the development of a series of detailed cognitive analysis and evaluation sheets (following the founding principle of the "Cultural Heritage of Basilicata"), which can be implemented over time (also thanks to the reference SIT), as an enrichment of the contents of the Plan.

A study such as the one carried out for the medieval site of Brienza, which concerns territories of outstanding historical-environmental values, not yet identified and recognized, could constitute a valid document of programmatic interpretation.

Given the historical importance of the site examined in the present research, which is also a territorial area linking the regions of Campania and Basilicata (which represent, in turn, one of the main communication node for the Tyrrhenian and Ionian cultures), we argue that our approach could constitute an important starting point for further new and challenging observations.

Aknowledgements. Thanks are due to the reviewers for the useful suggestions given.

References

Albert, C., Galler, C., Hermes, J., et al.: Applying ecosystem services indicators in landscape planning and management: the ES-in-Planning framework. Ecol. Ind. **61**, 100–113 (2016). https://doi.org/10.1016/j.ecolind.2015.03.029

Bartolucci, F., Peruzzi, L., Galasso, G., et al.: An updated checklist of the vascular flora native to Italy. Plant Biosyst. **152**, 179–303 (2018). https://doi.org/10.1080/11263504.2017.1419996

Daily, G.C.: Nature's Services: Societal Dependence on Natural Ecosystems. 1997th edn. Island Press, Washington, DC (1997)

Di Pietro, R., et al.: A survey of landscape planning in Italy, where application is utopia. An updated proposal for a shared landscape analysis model. Plant Sociol. **56**(2), 113–128 (2019). https://doi.org/10.7338/pls2019562/08

Doddato, V.: Relazione Generale. Piano di Assestamento forestale. Periodo di validità 2017–2026. Comune di Brienza, Brienza (PZ), Italy (2017)

Galasso, G., Conti, F., Peruzzi, L., et al.: An updated checklist of the vascular flora alien to Italy. Plant Biosyst. **152**, 556–592 (2018). https://doi.org/10.1080/11263504.2018.1441197

Giandon, P., Bortolami, P.: L'interpretazione delle analisi del terreno. Strumento per la sostenibilità ambientale, 1990th edn. Veneto Agricoltura - Azienda Regionale per i settori Agricolo, Forestale e Agro-Alimentare, Legnano (PD), Italy (1990)

Iamonico, D.: Multitemporal analysis of landscape of the Appia Antica Regional Park (Rome). Ital. J. Remote Sens. **40**(3), 27–37 (2008)

Iamonico, D., Di Pietro, R.: L'Ecologia vegetale per la conservazione attiva e la gestione sostenibile del territorio. Il caso dell'area protetta "Travertini Acque Albule" (Italia Centrale). Urbanistica **15**, 209–213 (2018)

Maraja, R., Barkmann, J., Tscharntke, T.: Perceptions of cultural ecosystem services from urban green. Ecosyst. Serv. **17**, 33–39 (2016). https://doi.org/10.1016/j.ecoser.2015.11.007

Milcu, A.I., Hanspach, J., Abson, D., Fischer, J.: Cultural ecosystem services: a literature review and prospects for future research. Ecol. Soc. **18** (2013). https://doi.org/10.5751/ES-05790-180344

Pignatti, S.: Flora d'Italia, 2018th edn. Edagricole, Bologna (2018)

Plieninger, T., Bieling, C., Fagerholm, N., et al.: The role of cultural ecosystem services in landscape management and planning. Curr. Opin. Environ. Sustain. **14**, 28–33 (2015). https://doi.org/10.1016/j.cosust.2015.02.006

Plieninger, T., Dijks, S., Oteros-Rozas, E., Bieling, C.: Assessing, mapping, and quantifying cultural ecosystem services at community level. Land Use Policy **33**, 118–129 (2013). https://doi.org/10.1016/j.landusepol.2012.12.013

Pedio, T.: La statistica Murattiana del Regno di Napoli, le condizioni economiche, artigianato e manifatture in Basilicata all'inizio del secolo XIX. Potenza, Italy (1964)

Pedio, T.: La Relazione Gaudioso sulla Basilicata (1736). Centro Librario, Bari (1965)

Stålhammar, S., Pedersen, E.: Recreational cultural ecosystem services: how do people describe the value? Ecosyst. Serv. **26**, 1–9 (2017). https://doi.org/10.1016/j.ecoser.2017.05.010

Thiers, B.: Index Herbariorum: A global directory of public herbaria and associated staff. New York Botanical Garden's Virtual Herbarium (2021). [Continuously updated]. http://sweetgum. nybg.org/science/ih/. Accessed 26 Aug 2021

Vasiljevic, N., Gavrilovic, S.: Cultural ecosystem services. In: Leal Filho, W., Azul, A., Brandli, L., Özuyar, P., Wall, T. (eds.) Life on Land. Encyclopedia of the UN Sustainable Development Goals, pp. 1–10. Springer, Cham (2019), https://doi.org/10.1007/978-3-319-71065-5_47-1

Modelling for Spatial Planning

How to Support the Decision-Making Process of Community Regeneration in Megacities Based on Evolutionary Game——A Case Study in Shanghai

Youmei Zhou, Junyan Lv, Zhihai Qu, and Daixin Dai[✉]

Tongji University, Siping Road 1239, Yangpu District, Shanghai, China
youmeizhou@icloud.com

Abstract. Urban regeneration is an inevitable trend at a certain stage of urbanization, and as the basic unit of the city, how to develop suitable regeneration participation mechanisms and strategies has become an important issue, especially for the comprehensive design and management of integrated strategies to promote internal drive so that it can operate on a regular basis. The problem is systemic and complex, with different stakeholder cooperation mechanisms driven by different actors in the government-led, market-led and co-production models of participation and community planner participation models. Game theory has gradually become an important tool for delineating and studying interest groups, but simulation of game processes and derivation of outcomes based on participation mechanisms have not yet been applied in community regeneration. Moreover, using only case studies will have a long practice cycle with large economic and social impacts, and lacks reasonable and efficient experimental support. This study uses an evolutionary game model to explore the impact and association of different types of community regeneration participation patterns under various benefit reward and punishment strategies. Based on this, a system dynamics model is developed to analyze the dynamic changes of the multi-party competition in typical situations with the help of data simulation, and to explore the promotion of the balance between the reward and punishment strategies under different participation patterns, to achieve the strategic optimization for sustainable development. The expected research results will not only enrich and improve the theory of urban regeneration, reveal the influence mechanism of the dynamic cooperation strategies of different stakeholders in regeneration, but also provide scientific basis and technical support for decision making in fine urban governance and organic development, thus this project has important theoretical and practical value.

Keywords: Community regeneration · Participation mechanisms · Evolutionary game theory · Simulation · Decision-making support

1 Introduction

The aim of this article to present a proposal for a method to construct sustainable participation machines of ageing appropriate renewal project in shanghai, an upgraded method

© The Author(s), under exclusive license to Springer Nature Switzerland AG 2022
D. La Rosa and R. Privitera (Eds.): INPUT 2021, LNCE 242, pp. 83–93, 2022.
https://doi.org/10.1007/978-3-030-96985-1_10

based on Evolutionary Game theory has been introduced to this research to investigate the dynamic simulated system of cooperating participation machines and reveal the different strategies for different type of communities. This paper is treated as the first step to further research on including in urban planning the topic of active reaction of elderly people. However, this paper divided into three main section to explain the significance, the aim and procedure and the mixed method design, and the expected innovations.

In this new period of development, Shanghai is constantly exploring ways and means of working to better respond to the new aspirations and needs of the people of Shanghai for a better life through three transformations: firstly, a people-oriented conceptual transformation that takes the needs of the people as the starting point and promotes the upgrading of planning and construction standards; secondly, a transformation of community governance through multi-dimensional collaboration to create a new pattern of governance for common construction, governance and sharing; thirdly, an all-round digital transformation that uses New digital technologies will empower community services and community governance. Community transformation and community upgrading will inevitably lead to community renewal, but further research is needed on the public participation mechanism and incentive strategies for community renewal, and to truly implement the organic renewal of "urban cells", the government still needs to formulate corresponding policy guidance and incentive strategies to promote its sustainable development.

Based on the positioning needs of these three types of community regeneration, this study is divided into three scenarios for analysis.

- Scenario 1: Age-friendly renewal

The Shanghai Municipal Bureau of Civil Affairs will complete the retrofitting of housing security and accessibility facilities for low-income and other disadvantaged elderly households from 2012 to 2019, and carry out a pilot programme of age-friendly retrofitting of the home environment for all elderly people from 2020 (hereinafter referred to as "age-friendly retrofitting"), with the National Office for the aging and other in October 2016, 25 ministries and commissions jointly issued the "Guidance on Promoting the Construction of a Livable Environment for the Elderly", which proposes to strengthen the accessibility of residential areas and meet the basic needs of the elderly for safe access. The enforcement and supervision of the regulations on the construction of a barrier-free environment will be strengthened, and the relevant standards for the construction of barrier-free facilities in new residential buildings will be strictly enforced. In the next step, the Shanghai Housing and Construction Commission, the Municipal Office for the Aging and other departments will focus on the barrier-free access and safety needs of the elderly, and promote the establishment of a regular mechanism for the construction and maintenance of barrier-free facilities in old residential areas. The issue of how to establish an organic and sustainable mechanism to meet the needs of multiple interests is an issue that needs to be addressed now.

In Shanghai, the optimization and renewal works for the elderly are progressing from point to point, from the home to the public space of the building to the overall external public space of the old district, but there is still a lack of research on the regular participation mechanism for the optimization of the elderly. While the increase in the scope of regeneration will increase the state's financial subsidies, how to better engage social capital, while maximizing the quality of life and well-being of elderly residents, balancing the need for all-age services in the community space and avoiding 'gentrification' is the next phase of community ageing optimisation.

In this context, older communities are often a pain point for urban regeneration, as they are often absent from the scope of municipal redevelopment and lack of economic drivers. As these communities are characterised by a significant amount of older residents, the question of how to develop a sustainable and normalized participatory mechanism for the regeneration of older communities has become an important issue, and how to set up reward strategies to guide and drive the collaboration of residents, planners, the market and the government in order to form a normalized model of organic regeneration in response to the aging process.

- Scenario 2: High-quality public space in a community built for sharing and cooperation

Shanghai's latest Shanghai Urban Master Plan (2015–2040) outlines the need to change the concept of urban development, clearly proposing "zero growth" or even "minus growth" in construction land. Shanghai's urban development has entered a stage where urban regeneration is the main line, i.e. promoting urban regeneration to a stock planning; aiming at improving the vitality and quality of the city, actively exploring a progressive and sustainable organic regeneration model, meeting the spatial needs of the city's future development by renewing the use of the stock of land, and at the same time preserving and inheriting the city's culture, forcing the way of land use to change from outward and rough expansion to inward growth. It will also promote the transformation of spatial use into an intensive, compact, functional, low-carbon and efficient one. This also lays down the overall trend of China's future urban development to fine tune the goals of construction and administration.

In early 2015, the Shanghai city government announced the Three-Year Action Plan for Strengthening Comprehensive Governance of Residential Communities in Shanghai (2015–2017), which demands to promote the formation of a four-in-one and positive interaction pattern of comprehensive governance of residential communities with government supervision, market-led, social participation and residents' autonomy, gradually shifting from top-down government management to bottom-up and top-down integrated urban governance. In such a context, old communities often become a pain point for urban regeneration because they are not covered by municipal resettlement and lack of economically driven improvements. However, the traditional design participation model of bottom-up social participation and residents' autonomy suffers from the difficulty of expressing their demands in a timely manner, and the lack of participation in regeneration feedback. In addition, investors' expectations of profit in regeneration often lead to higher land prices due to space upgrading, resulting in the "gentrification" of land use and the unrealisation of true sharing. At present, there is a lack of in-depth research on the division of interests and responsibilities and the mechanisms of participation, and there is a lack of reliable theoretical basis and data to support the decision-making process of enhancing cooperation and participation. The key issue is how to improve the health and quality of life of residents in the renewal of the built environment. Appropriate methods and incentive strategies can guide and enhance the public's awareness and behaviour of participation, thus enhancing social vitality and community identity.

- Scenario 3: Digital transformation and renewal of communities

With the development of the times and data technology and the trend of digital transformation of cities, we need to promote the wide application of digital technology in public services and urban administration. We will apply digital technology to solve social and public problems, develop various applications for the people, build a new type of smart city, build an intelligent governance system based on the "city brain", strengthen the application of digital technology in urban planning, construction, governance and services, and enhance the scientific, refined and intelligent level of urban management. As a basic unit of urban development, communities are closely related to the quality of people's lives, and their digital development and intelligent management are of great importance in supporting the development of digital cities. The digital transformation of communities is not only about enhancing the community's ubiquitous data-awareness technology, i.e. the physical attributes of space, but should also focus on the sensing of social aspects and residents' emotions. However, existing research and practice has focused more on physical space and less on the latter two, and the relationships between the three in action. In the creation of spatial digital optimal design and governance, it is easy to produce "technology oligarchy", so that some residents because of differences in background or conditions become the disadvantaged groups of digital communities, and the public participation mechanism of which still needs to be clarified, and this will also affect the empowering utility of the residents' management and participation by means of digital instruments by the authorities. This will be a complex issue involving multiple rights and new challenges in new pathways.

1.1 Game Theory and the Urban Regeneration

Urban renewal has long been an important part of the urban policies of city governments in both developing and developed countries, particularly with regard to the upgrading of old cities and urban environments. Over the past few decades, and even throughout the last century, the trend towards urban renewal has continued unabated, starting with slum clearance projects in the United States in the early 20th century, followed by inner-city redevelopment to improve living conditions in some European countries after the war. (Gilbert 2009) Currently, urban renewal projects of various scales are being vigorously implemented in developing countries such as China and Turkey, as well as in developed countries such as the United Kingdom and Spain, as a means of revitalizing local economies. In research, any stakeholder affected by organizational decisions and actions is referred to as a Stakeholder (Stakeholder) and urban regeneration decision making is essentially a multi-criteria decision making problem with multiple stakeholders (Trutnevyte et al. 2012), and decisions will be made under multiple criteria representing different objectives. (De Brucker et al. 2013) Since the relative importance of each criterion may not be the same, many methods have been proposed to deal with multi-criteria decision-making problems with different decision weights (Chu et al. 2020), such as analytic hierarchy process, entropy weight method and fuzzy set approach. And stakeholders from different backgrounds have quite different views on the importance of decision criteria for urban renewal, and also, since the decision process of urban renewal is not done at once, it is a long-term dynamic process including the interaction among stakeholders. (Wang et al. 2014) As a result, alternative approaches to these urban renewal decision-making problems have been proposed from different perspectives, including game theory. (Abdalla 2016) The analysis of policies from a multi-stakeholder perspective is a basis that has been widely recognized and applied by scholars in the field of social economics, both domestically and internationally. (Barari et al. 2012; Chatterjee et al. 2012) While it has been less applied in urban renewal, it has not been thoroughly discussed in the study of the renewal of old neighborhoods, especially in the study of age-appropriate optimal renewal, especially the simulation of mechanisms in the framework of evolutionary games.

In general, the main stakeholders involved in urban regeneration are those who can intervene in the project implementation during the construction or operation of the project, and they mainly include the government, developers, and residents. (Liu 2020) In the general hypothesis, the core interests of the three parties (i.e., government, developers, and residents) are highly dispersed and do not overlap sufficiently, with the government playing a dominant role, representing mainly the public interest and its main efforts to stimulate socio-economic development from a macro perspective; developers are mainly concerned with direct or indirect benefits (especially the former) and projecting a positive corporate image. (Wang et al. 2019) In the study of community regeneration, bottom-up regeneration model and demand extraction are gradually paid attention to by the government, while diversified participation systems such as community planner system and community residents' participation in design workshops emerge, but there is still an internal game of interests for the interest group of residents, i.e. the diversity of residents' needs, and what is particularly prominent nowadays is the balance between the needs of the elderly and other For example, in the current community renewal, many elderly people have contradictions with young people in the proportion of social space, walking space, green space and parking space, and in the renewal of elevator access, there are also objections from young people in terms of cost and space utilization, and these are closely related to the needs and perceptions of the elderly and other residents. Other residents' needs and perceptions are closely related to determine the post-renewal expectations and impacts, such as sense of belonging, local identity, and social cohesion, so the overall benefit division and decision making is a complex and integrated multi-level model, and evidence-based studies considering this holistic perspective are still lacking, and the reward strategies in the participation mechanism have not been thoroughly analyzed.

2   Research Procedure and Methods Design

2.1   Research Aim

This study selects Shanghai, the city with the largest resident population and the highest degree of aging, as a representative case study of the aging response strategy of megacities. Through the identification of stakeholders and influencing factors, an evolutionary game model will be constructed and simulated for different types of communities in Shanghai, and a balanced sustainability solution will be sought to maximize the interests of all parties by adjusting the incentive strategy, to enhance public participation and improve the effectiveness of renewal in physical, social and economic dimensions (Fig. 1). The study will propose the optimization of age-appropriate renewal mechanisms and renewal incentive strategies for different models (community types) to achieve sustainable organic renewal.

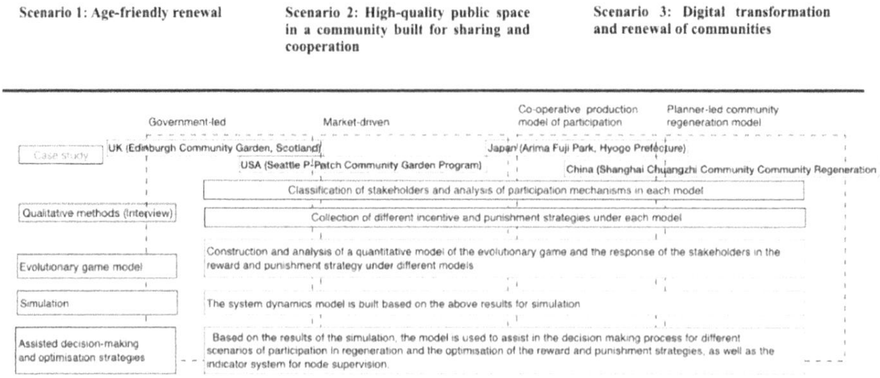

Scenario 1: Age-friendly renewal	Scenario 2: High-quality public space in a community built for sharing and cooperation	Scenario 3: Digital transformation and renewal of communities

Fig. 1. The conceptual framework of research design

2.2 Research Objectives and Content

1. Summarize participatory mechanisms and incentive strategies for age-friendly renewal of the built environment in the community

   Drawing on the theories and experiences of international urban space, community age-friendly optimal renewal, and the creation of age-friendly city renewal construction, summarize the multidimensional demand framework of the elderly for the environment and the existing theoretical response and participation mechanism of environmental renewal, sort out the goals, mechanisms and strategies of age-friendly optimal renewal in China; through analyzing the evolution of the theory of age-friendly city and the influence of mechanisms and strategies in the cases of age-friendly optimal renewal, and from the stakeholder Through the analysis of the evolution of aging-friendly city theories and the influence of mechanisms and strategies in the cases of aging-friendly renewal, we clarify the current problems of different types of community aging-friendly renewal and the influencing factors of participation mechanisms, and build a theoretical framework to provide a new development path for the comprehensive and comprehensive promotion of aging-friendly renewal.

2. Refined analysis of each stakeholder coupled with all-age multidimensional needs

   Through document review and case study, the main types of mechanisms and strategies in community ageing renewal are clarified, and the impact paths of participation mechanisms and incentive strategies in different types of community ageing renewal are clarified by investigating the dynamic behavioral processes of multiple stakeholders in community ageing renewal and their social, economic, and resident life and psychological impacts. Through literature and case field interviews, the spatial, perceptual, and social needs of each age group are summarized based on rooting theory, and all-age needs are coupled to delineate common needs and non-cooperative needs, and then the refined needs framework and spatial design guidelines for all-age residents in different types of community contexts are proposed (Fig. 2).

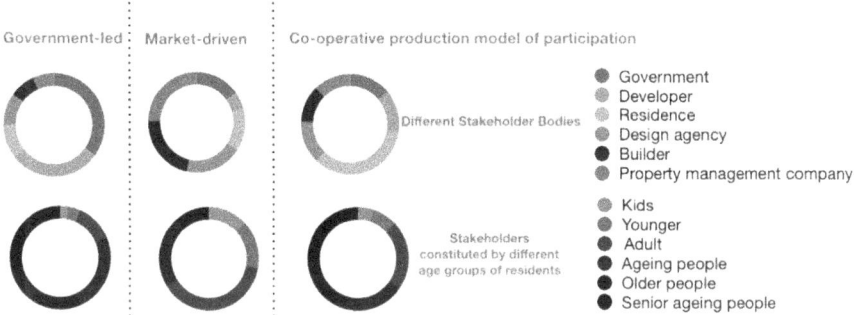

Fig. 2. Multiple stakeholders in community ageing renewal

3. Constructing an evolutionary game model of community participation mechanisms for age-friendly renewal

Based on the above research results, we introduce Evolutionary Game Theory to delineate the stakeholders in different community types, construct an evolutionary game model to analyze how different existing incentive strategies improve the cooperation among stakeholders, explore the response of stakeholders to strategies under different participation mechanisms, and simulate the basic mechanism of each group to update their strategies in each interaction. The basic mechanism of adaptive strategies is simulated, and each group changes their strategies through reward and punishment, and other objects "comparing" and "learning", introducing micro-psychological Adaptation-level theory to add theoretical understanding to the process of adaptation and change. The adaptation-level theory is introduced to add theoretical understanding to the process of adaptation and change, and how to finally reach a long-term sustainable state, i.e., a "Nash equilibrium" to maximize the benefits of the group. In this process, this study will propose a basic model of evolutionary game applicable to the scenario of ageing regeneration and optimize the original model by collecting and analyzing the data of variables in the case, and propose an optimization model applicable to different community types and participation mechanisms to achieve the demand of fine governance of urban development.

4. Suggesting optimization and incentive strategies based on "Nash equilibrium" for different old-age renewal mechanisms

Based on the results of the above analysis, we project the existing policies and propose the optimization of the renewal participation mechanism and incentive strategies for multiple parties for the next comprehensive implementation of the community built environment to meet the basic living and safety needs of the elderly in Shanghai, and implement the normalized and organic development of age-appropriate renewal; we propose the proposal of the multi-dimensional needs of the elderly and the "all-age friendly" The proposal is oriented to the dual goal of "all-age friendly" (Li 2019) which is to meet the higher Maslow's needs level of community spatial renewal and environmental perception, social dimension enhancement, to provide a basis for the government's strategic deployment to gradually alleviate aging. It also provides a scenario framework of community subunits for building smart cities and realizing digital twins.

2.3   Research Methodology Design

The study integrates the multidisciplinary cross-fertilization theoretical constructs of urban science, psychology, gerontology, and socioeconomics, blends the evolutionary game modeling wheel of rational economics with the adaptive water assessment theory of micro-psychology, and proposes a participatory mechanism for optimizing and stimulating the age-friendly renewal of community built environments under the goal of sustainable organic renewal and all-age friendliness from the perspective of sustainable organic renewal responding to the national needs of aging and digitalization. scientific issues, constructing theories and generalizing existing mechanisms-stakeholder and demand variables identification-evolutionary game modeling and simulation operations-policy projection and decision support, respectively, applying systematic literature review method and case study method, interview method, field research, rooted theory qualitative analysis method, quantitative parametric modeling methods for refined research.

1.   Systematic literature review method and case study method

The systematic literature review method is used in the first research content, through 5 steps, i.e. defining research objectives, selecting database, identifying keywords, selecting compatible articles, and finally extracting data, to sort out the development lineage, impact mechanism, and incentive strategy of elderly-friendly renewal for this study, and to build a theoretical foundation; through the case study analysis method, the completed elderly-friendly renewal projects and ongoing projects will be analyzed in depth with web data, literature, and field survey, and the impact and effect of the existing mechanism will be summarized. Through the case study method, we will analyze the impact and effect of the participation mechanism and application strategy, and summarize the existing mechanism, and the field research will combine the interview method and network opinion data analysis method to make a real and comprehensive evaluation of the impact and effect, to compare and analyze the advantages and disadvantages according to different situations.

2.   Interview method and rooted theory qualitative analysis

Grounded theory for qualitative research has a wide range of applications in sociology, communication, and psychology, and is used to develop theory through the logical refinement of original vivid and real case data and then outline and abstract new concepts. The application in this study is in two aspects: firstly, the investigation of the dynamic behavioral process of stakeholders and its social, economic, residents' life and psychological impact, and the clarification of the influence path of the participation mechanism and incentive strategy of different types of community age-appropriate renewal; secondly, the identification of the multidimensional needs and interest elements of residents in each age group through interview data. Based on the coding of the interview content using NVivo qualitative data analysis software, there are six types of coding methods in the fifth edition of Sociological Research Methods, and this study will use three of them according to the research objectives: "open coding" (open coding), "spindle coding" (spindle coding), and "coding of the community" (open coding). "This study will analyze the interview data using three of these coding methods: open coding, thematic coding, and focused coding, and coupling the results of the data analysis of the thematic coding.

3. Quantitative Parametric Evolutionary Game Analysis Method-Modeling -Optimization-Simulation-Nash Equilibrium

Evolutionary game theory specializes in finding stable strategies in an evolving population and tracking evolutionary patterns in a long-term dynamic environment, its ability to overcome the shortcomings of traditional games, especially in problems related to strategic decisions, is widely used to reveal dynamic interactions among stakeholders, has been successfully applied to predict strategic decisions of stakeholders, and has developed a number of games that can positively induce incentives or penalties. Based on the main results of the first two research elements, the data-based model optimization is performed for different types and scenarios, and the simplified analysis steps of the basic model are shown below.

Use $x_{i1}(t)$ Indicates that the first layer is in the t time to select the policy i. The percentage of the Use $u(i1, x_{i1})$ denotes the proportion of game participants choosing a strategy in the x_{i1} state choose the strategy i1 of fitness (benefit, payoff). And use $u(x_{i1}, x_{i1})$ denotes the current strategy state x_{i1} the average comfort level of the.

For each layer of the game the proportion of players choosing strategies different strategies (e.g. accepting the policy) are x_{i1}, y_{j1}, z_{k1}. x_{i2}, y_{j2}, z_{k2}. (Note, here there are not necessarily 3 game objects in each layer, there may be only two in some layers and 4 in others, selected in the study based on the findings of study I and II).

The change in the proportion of players at different levels of the evolutionary game can be obtained by modeling the benefit (payoff) of the game participants. That is, for each individual will choose the decision that benefits him or her more by his or her current decision versus the overall average decision. That is, the dynamics can be obtained by the difference between the two. For the bottom.

$$\frac{dx_{i1}}{dt} = x_{i1} \left[u(i1, x_{i1}) - u(x_{i1}, x_{i1}) \right]$$

The above $u(i1, x_{i1})$ the lesson is seen as the first tier of participants.

Similarly, for the second tier

$$\frac{dx_{i2}}{dt} = x_{i2}\left[u(i2, x_{i2}) - u(x_{i2}, x_{i2})\right] + \frac{dx_{i2}}{dx_{i1}}\frac{dx_{i1}}{dt}$$

Here the $\frac{dx_{i2}}{dx_{i1}}$ indicates the coupling between the first and second layers. The final $\frac{dx_{i1}}{dt}$ is then calculated at the first layer, and the latter item shows the coupling between the two layers and the extent to which the bottom layer affects the top layer.

3   Conclusion

This research is based on theoretical frontiers and the urgent needs of the times, and gives a sustainable perspective on urban regeneration, especially community regeneration, and offers a practical approach to digitalization and intelligence, giving a new perspective on collaborative mechanisms for community regeneration and providing powerful research tools and technical implementation in the future.

The main idea of this study is to design a suitable incentive strategy to improve the effectiveness of the renewal mechanism and maximize the interests of all parties, so as to achieve the purpose of normalizing and organizing the development of ageing-friendly renewal. This study uses evolutionary game theory to simplify and shorten the research period by simulating the trial and error process, and the response to the incentive strategy is more general and provides an important basis for government decision.

Acknowledgment. This paper is an output of Research and Development Project of the Ministry of Housing and Urban-Rural Development "Research on the Development and Planning Application of the Intelligent Platform of Full Volume Data and Scenario Simulation for Smart Communities under the Concept of Social Health and 'Digital Twin'" (2021-K-148).

References

Abdalla, S.S., Elariane, S.A., El Defrawi, S.H.: Decision-making tool for participatory urban planning and development: residents' preferences of their built environment. J. Urban Plan. Dev. **142**(1), 04015011 (2016)

Barari, S., Agarwal, G., Zhang, W.C., Mahanty, B., Tiwari, M.K.: A decision framework for the analysis of green supply chain contracts: an evolutionary game approach. Expert Syst. Appl. **39**(3), 2965–2976 (2012)

Chatterjee, K., Zufferey, D., Nowak, M.A.: Evolutionary game dynamics in populations with different learners. J. Theor. Biol. **301**, 161–173 (2012)

Cho, C.J.: An analysis of the housing redevelopment process in Korea through the lens of the transaction cost framework. Urban Stud. **48**(7), 1477–1501 (2011)

Chu, X., Shi, Z., Yang, L., Guo, S.: Evolutionary game analysis on improving collaboration in sustainable urban regeneration: a multiple-stakeholder perspective. J. Urban Plan. Dev. **146**(4), 04020046 (2020)

De Brucker, K., Macharis, C., Verbeke, A.: Multi-criteria analysis and the resolution of sustainable development dilemmas: a stakeholder management approach. Eur. J. Oper. Res. **224**(1), 122–131 (2013)

Gilbert, P.: Social stakes of urban renewal: recent French housing policy. Build. Res. Inf. **37**(5–6), 638–648 (2009)

Li, S.Y.: Inclusive design-a public space renewal strategy for all-age community goals. Urban Dev. Res. **26**(11), 27–31 (2019)

Liu, Z.: Evaluation of Urban Settlement Renewal Needs and Planning Response from the Perspective of Multiple Subjects. Northwestern University (2020)

Trutnevyte, E., Stauffacher, M., Scholz, R.W.: Linking stakeholder visions with resource allocation scenarios and multi-criteria assessment. Eur. J. Oper. Res. **219**(3), 762–772 (2012)

Wang, H., Shen, Q., Tang, B.S., Lu, C., Peng, Y., Tang, L.: A framework of decision-making factors and supporting information for facilitating sustainable site planning in urban renewal projects. Cities **40**, 44–55 (2014)

Wang, X., Yang, L., Ye, Y.: Contention for urban state space and the rise of society in China: a case study of "housing planting" in Hohhot. Cities **92**, 219–229 (2019)

Urban Transformations for the Transit City. An Integrated Approach for a Medium Sized Mediterranean City

Luca Barbarossa[✉] [iD]

Department of Civil Engineering and Architecture, University of Catania, Catania, Italy
luca.barbarossa@darc.unict.it

Abstract. The most innovative research contributions, both in the field of urban and transport studies, proposes a plenty integrated approach to land use and transport planning. In particular, several researches deal with the relationship between the settlement forms and mobility behaviors to provide an acceptable level of accessibility and mobility for the inhabitants.

Starting from these contributions this paper briefly argues about the need to address urban and mobility planning, toward to an integrated approach, assuming a specific point of view oriented to locate mixed use functions along transit corridors.

This perspective is discussed here also referring to the case study of Catania, a medium sized metropolitan city, in Southern Italy, were recent investments on the railway networks, as well as the ongoing revision of the city land use plan, are changing considerably the mobility of the entire metropolitan area.

The contribution focuses on the adopted planning criteria with the aim to activate new patterns for land use and transport integration in order to reduced private car dependence. The challenge of the proposal presented is to define a set of specific rules that can virtuous urban transformations by contextualizing these general and widely accepted principles to local conditions.

1 Land Use and Transport Integration. A New Opportunity for the Sustainable City

The relevant evolution of metropolization process, during the last decades radically changed the urban structure and the efforts that characterize contemporary metropolitan cities. Among this an adequate public transport system and a right accessibility level are considered significant issues as they are part of citizens right. (Secchi 2010; Appleyard and Forst 2019). As a matter of fact, reducing mobility and accessibility within the city, means also produce unwanted social effect such as marginality and segregation (Hickman et al. 2015).

Although the relevance of the question, in Italy, sustainable urban mobility is still an unsolved issue in a number of metropolitan cities due the lack of effective policies and planning tools focusing on the integration and fully correspondence between public transport strategies and land-use plans (Curtis 2015).

© The Author(s), under exclusive license to Springer Nature Switzerland AG 2022
D. La Rosa and R. Privitera (Eds.): INPUT 2021, LNCE 242, pp. 94–103, 2022.
https://doi.org/10.1007/978-3-030-96985-1_11

Despite the number of solutions experienced in many European cities, in Italy urban and mobility plans seems to be affected by a huge delay regarding integration policies. An unsolved issue that generates "*the genetic abnormality*" characterizing the growth process in most of the Italian cities (Campos Venuti 1999).

Urban and transport plans whose integration rarely goes behind rhetorical announcements, are often implemented trough planning processes carried out by different actors, following different rules and techniques, as well as different endorsement, timing and funding channels.

The result is a deep lack of effective public means of transports, integrated with the urban settlements, particularly in South regions, whose transport systems, almost based on private cars, caused unsustainable congestion levels.

On the contrary, looking at the contemporary city, with the aim to address urban growth and physical transformation toward sustainable models, require a planning approach oriented to provide mainly the employment of public means of transport, integrated with the urban context.

According to this theory, transportation and urban planning must be faced trough structural and action plans aimed to fight developing models based on private cars that seems to be increasingly conflicting with the contemporary city dynamics.

Several studies (Marshall and Banister 2007; Kenworthy 2010; Hickman and Banister 2014) have focused on various aspects of integration policies, highlighting the role of land use patterns in improving accessibility and increasing the use non-motorized means of transport.

An important role in shorting private cars use in urban areas is attributed to transit corridors located in strategic places among urban settlements. Around them urban plans should define a set of specific rules and widely accepted principles such as concentrate developments, characterized by high density and mix use, along public transport nodes, protecting of open space and nature (La Greca and Martinico 2011).

Plans, both at the urban and regional scale, can introduce new contents, features and design criteria oriented to address land use policies toward a plenty integration with mobility choices and evolve land use zoning according to new mobility spaces and flows (Curtis 2015).

These topics are clearly referred to *Transit Oriented Development (TOD)* approach (Calthorpe 1993; Cervero et al. 2002), aim at creating new settlements, along the transit corridors, characterized by mix residential, retail, office, open space, and public uses in a walkable environment, making it convenient for residents and employees to travel by transit, bicycle, foot, instead of cars. In this scenario transit stops became the new urban polarity were developing model are based on integration and interaction between land use and public transit systems (Curtis 2015).

Finally, TOD is designed in order to create high density, compact and mixed urban form and to curbing sprawl and low-density urban models with the related private car dependence[1] (Deakin 2019).

2 Mobility, Accessibility and Strategic Choices for Catania Master Plan

The themes briefly outlined above, are particularly interesting if referred to a specific experience of urban and transport planning currently under way in Catania, a medium size metropolitan city located in South Italy.

Catania is the most important center of eastern Sicily, extending its influence well beyond the administrative borders of its Province. The city is the focus of a large conurbation that stretches, almost continuously, along the eastern coast of the island, from Messina to Syracuse. The main city extends $180,90$ km^2 and population is 300.356, according to the last census (ISTAT 2021).

From the first half of the 20th century, the city has increased its role of marketplace and supplier of services for the entire South-Eastern and Central Sicily. The favorable location, along the coast, well connected to motorways and railways and the presence of a commercial port and of a busy airport gives to the city a strategic role within Sicily (La Greca and Martinico 2018).

The current settlement has developed around the historical area, extending mainly toward north and north east along the coast, reaching existing agricultural and fishing villages beyond the city administrative borders. The result is a highly urbanized settlement with a high percentage of urban sprawl. Today, about 60% of the total population lived outside the main city. Encouraged by the spread of private cars and low real estate prices, this low-density urbanization has both generated serious traffic congestion and badly affected the fragile Mount Etna traditional rural landscape.

Today the conurbation includes 19 municipality that have intense functional relations with the main city, forming a continuous urban fabric where the only difference is

---

[1] Among many relevant experiences found in new residential districts built during the last decades in some North European cities, can be useful underline the cases of Vauban and Reisefeld, two residential districts in Freiburg, designed according transit-oriented principles, and structured around a new public transport system, which ensures high accessibility levels for the inhabitants. Hammarby in Stockholm and Ørestad, Copenhagen are other new peripheral district, known as important experience of environmental urban settlement. The districts planned and designed, according integrated land use and transport policies and following transit-oriented principles, represent an important and successful contemporary example of planned sustainable TOD. Significant investments have been made in public transport in the entire district, both in the form of the new light rail and bus traffic. These urban transportation policies, mainly oriented to public transport use, have significantly reduced emissions related to transport (Inveroth et al. 2013; Knowles 2012).

political jurisdiction. It is essentially an "automobile city" for about 700.000 inhabitants, with a very limited presence of public transports network and other public services[2].

In addition, the metropolitan city is inadequate to cope with traffic congestion, negative phenomena widespread in the entire metropolitan area, increased by the shortage of effective public means of transport[3]. As a consequence, mobility represents one of the most important issue to address in order to improve urban sustainability and give a substantial contribution to mitigation and adaptation at climate change in urban area (Barbarossa et al. 2018).

After decades of inaction, the city is now facing a big and unrepeatable opportunity due the ongoing transformation of two rail systems, that could change strongly, not only the metropolitan rail transports, but also the entire mobility system as well as some important parts of the urban structure. At the same time, after many attempts, since 2019 the municipality started the review process of the city master plan, aimed at addressing some key critical issues for the city, such as the high level of congestion of mobility system as well as the relevant shortage of public spaces and services especially green areas.

Concerning mobility, and in particular rail transports, the main under way project is the underground, developed as the upgrade of a local narrow-gauge railway, built in 1885, connecting Catania with several towns around Mount Etna. When completed the line will become an efficient north–west connection, 30 km long, linking the city center with main social housing neighborhoods both north-west and south of the city center, with a terminus at the international airport[4].

Another important transport infrastructure is the upgrading of the urban and metropolitan section of the national rail alignment. It includes the construction of a new tunnel and four new urban stations that will allow the provision of metropolitan and sub regional services. The impact of this project on the city will be considerable, especially on the city waterfront, today occupied by rail tracks. The removal of this obstacle will open up to innovative urban solutions aimed at reconnecting the city with the sea in an area of a huge scenic value.

Despite this ongoing project will configure a new and better order of the entire metropolitan transport system, the lack of integration with urban planning, that characterize these infrastructure projects, would have negative effects on the city.

---

[2] The majority of trips from outside the main city (87%) are made by car. Public transport (13% of total mobility) is constituted mainly by low frequency, unreliable bus services (50% urban buses and 50% long distance buses), rail share is negligible. (Source: City of Catania, General Plan of Urban Traffic (PGTU), available at https://www.comune.catania.it).

[3] Euromobility 2021, the annual sustainable mobility national report, ranked Catania at the last place, among 50 main Italian cities (population up to 100.000). The more relevant data are related to the highest private cars rank, (with 77 cars per capita Catania is the first Italian city in a car density rank, compared with the national average of 60,5 and the European average of 52,4) the lack of car and bike sharing policies, the shortage of pedestrian areas and traffic calming measures (Euromobility 2021).

[4] Today, only 8.8 km with 10 stations are operating and 4.00 km and 5 stations are under construction. According the future developing programs, in 2027 the line will be developed for further 18 km and 12 stations. (Source: https://catania.mobilita.org).

As a matter of fact, the absence of integration policies, not only could have a negative effect in term of accessibility, but also could reduce urban quality of the entire city. In detail, in many cases rail stations not appear to have relationships with the urban context, a proper level of accessibility, and above all, they not play the role of morphogenetic elements for a new city development model.

However, it's clear that the transformation and upgrading of two important metropolitan rail systems, represent a big opportunity for the city and for the conurbation as whole.

In addition, the combination of the rail upgrading with the revision of the land use plan could became an occasion for implements strategies of the integration between the ongoing infrastructural project, and the new land use policy, that should become one of the most characterizing strategy of the new city master plan. Those strategies can contribute to trigger deep transformations within city, not only in terms of urban form, but also in terms of mobility and accessibility.

As a matter of fact, one of the key contents of the new master plan is referred to the empowerment of public transportation as well as the integration between land use and transport. According to the plan strategies, future land use around existing and new metro stations will take into account the new infrastructure.

According to these contents the new city master plan, should propose some strategies in order develop transit-oriented theories trough the concentration of new developments areas near the nodes of the on-going transit projects.

In this perspective the development of several brownfields located inside or at the edge of the city represent a huge opportunity to address planning choices toward TOD principles.

These areas identified according to accessibility criteria, presence of existing urban services, distance to the transit stops, will be developed following detailed plans, included in the city master plan, to be implemented by landowners on the basis of a Transfer Development Rights Program.

The development model should be based on a fully integration with transport system, reaching also a high-quality level of the new settlements, trough the identifying of new functions concentrated around public transit stops in a mix of residential, offices, retail, public services and green areas. The plan choices should also take into account accessibility criteria, particularly referred to pedestrian and cycling mobility.

Following this approach, transit stops will become the new urban polarities, interconnected and distinguishable among the urban fabrics, as a physical expression of a new transit city.

3   City Master Plan and Metro Rail Network Implementation

Most of the city strategies concerning mobility, are oriented to increase the efficiency of the rail system and at the same time, to reduce some critical aspects of the upgrading project, carried out by the national rail company.

The existing urban structure suggest, as a better solution, a rail link which would have a dual effect of upgrade the urban rail line and create new accessibility to part of the city that could be interested by urban renewal projects.

The proposal consists in a new railway gallery, connecting the central station with a new urban station south located, near the international airport. This new railway track, will pass through the south–west located historical districts of the city, and the upgrading of the line will include two new stations, as well as the renewal of an existing station and the reuse of a freight station, converted in urban one (Fig. 1).

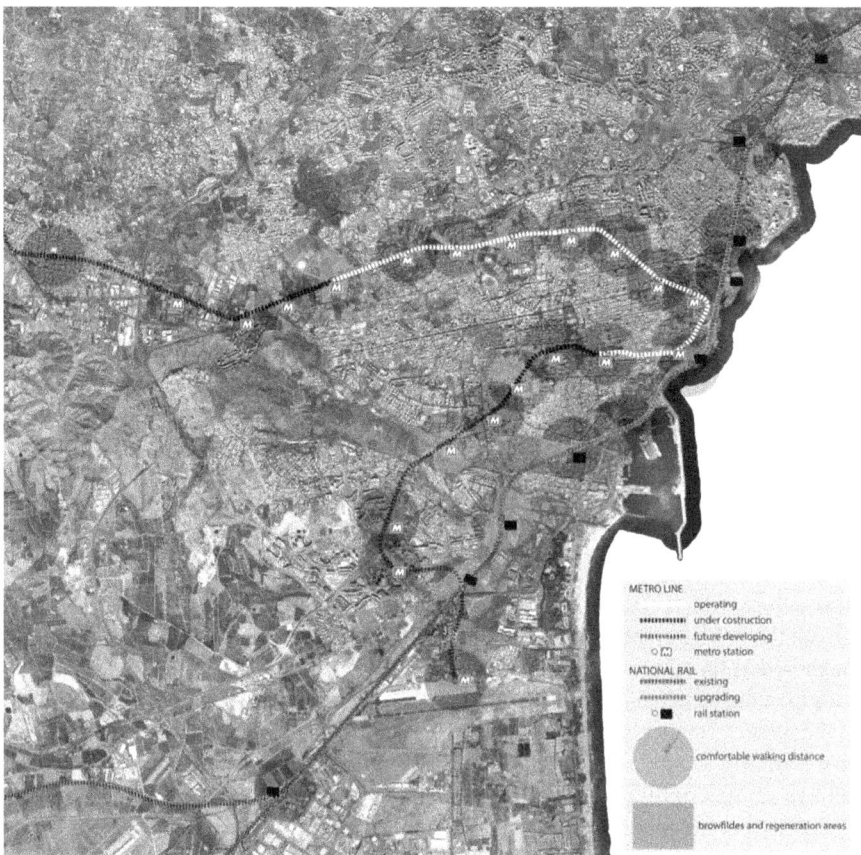

Fig. 1. Metro line and National Rail alignments and developing areas

The new rail link includes many qualifying features such as the implementation of a passenger's rail services at the metropolitan scale (metro rail), operating in a large part of the urban area.

In addition, the new metro rail includes several stops in areas not served by public transports, were, according to the reviewing city master plan, will be implemented urban renewal programs, characterized by, densification, mixing uses and public spaces.

The project will also permit the reuse of 20 hectares of coastal areas, today occupied be the rail tracks. This area, according to the new plan strategies, could be convert in office, residential, retail with a big amount (roughly 16 hectares) of urban public space.

The new destinations would generate the economic resource useful to the develop of railroad upgrading as well as the new public spaces.

Other aspects are related to the achievement of urban renewal intervention in some historical districts, characterized by huge decay phenomena.

Moreover, a new high capacity and high frequency metro train service, not only will contribute to reduce vehicular congestion in the area but also will provide to connect the district with the inner city and the international airport. As a consequence of new accessibility, the historical district, after a reshaping and a renewal process, could play a new role, of touristic and services area related both to the airport and to the city.

Same consideration could be also done regarding the other railway infrastructures located in the south part of the city. In detail more than 25 hectares of freight train station and rail maintenance depot abandoned or underutilized, could be reutilized as new urban areas were the plan will provide to allocate new valuable urban functions.

The implementation of the new metro rail will allow the reorganization of public spaces, around the new stations.

For example, the planned station designed inside the port area, close to the historical city, would become a big opportunity for a redevelopment project of the area in order to create new urban public spaces and localize a new gate for the most valuable a liveable part of the historical city.

Furthermore, the tracks removal by the actual line, elevated above the street, would allow the transformation of the viaduct in an elevated linear public space, as happened in other cities all over the word. This approach would permit the start of a renewal process useful to address the actual degradation phenomena caused by the existing railway built at the end of the 19th century.

The proposed strategy could present some critical aspect such as the high costs of infrastructure, which needs a huge amount of public investments, as well as the difficult in create the public private partnership essential to develop the project, but usually not enough widespread among local investors.

Despite the weakness of the proposal, mainly related to the implementation difficulties, the project, focusing on the upgrading of public rail transport, and as a consequence, on accessibility to the historical district, represent a big opportunity for the entire city. The huge amount of investments will be balanced by the increase of value of the transit station area and the renewal historical districts interested by the plan.

3.1 Planning Around Transit Stations

Among the city a number of areas, considered as strategic for the new development plans and interested by the metro line, were chosen as case studies. The chosen areas present different characteristics in term of location, density, land use, public space rate.

In order to identify the most effective strategies for the development or regeneration of the areas, some detailed analysis was carried out concerning land use, population, accessibility[5]. Land uses were mapped with GIS by using recent vector cartography and orthophotos, population data were obtained from National Census.

---

[5] The land use maps are one of the results of the research activity conducted, from 2001, by LAPTA (Laboratory for Territorial and Environmental Planning) at University of Catania (http://www.lapta.dicar.unict.it).

The analysis of current land uses shows, in several sites, low values of mixed uses around the new metro stations. Residential uses are prevalent in the city centre, with a lack of public spaces, services, retail.

Pedestrian accessibility was also evaluated trough an analysis of the preferred pedestrian paths and assuming an average speed of 4 km/h and a maximum time of walking (10 min). Also emerges from the analysis a low level of accessibility, also related to road safety and slopes.

The results of the analysis confirm that the new metro stations were located only according to engineering considerations, without any attention for pedestrian accessibility, land uses or population (Table 1).

Table 1. Land use around the metro station (500 mt)

Station	Residential %	Mixed %	Commercial %	Service %	Undeveloped %	Other %
Milo	56.1	9.4	0.0	24.8	3.4	6.3
Cibali	70.8	2.5	1.3	10.9	13.7	0.9
S. Nullo	74.1	6.7	1.1	6.7	14.1	1.4
Nesima	46.2	0.4	0.5	4.7	42.9	5.3
Fontana	20.3	5.0	7.6	11.9	41.8	13.4
Monte Pò	15.3	0.0	22.5	8.6	29.6	17.5
Misterbianco	0.5	0.0	53.2	0.0	8.7	37.7

The result of the analysis carried out on the study areas, confirm that a carefully land use plan is absolutely necessary in order to develop effective transit policies for the metropolitan area.

As a matter of fact, the developing and upgrade of major urban infrastructures, such as rail network, is an occurrence strictly related to urban development processes, that can definitely change urban dynamics in terms of land uses, spatial relationships, economies.

Concerning the study areas, new development, regeneration or infill should be operated with the aim to concentrate residential settlements and other functions around transit nodes in a renovated urban environment that can improve accessibility to the stations and encourage non-motorized means of transport.

In this perspective urban transformations should be oriented to rebalance mixed land uses, reorganize and increase the amount of public space, services, create an urban environment that encourage walking and cycling. Furthermore, the large availability of brownfields or underused land around many stations can be considered an opportunity for transformation such as redevelopment or infill.

4 Conclusion

The present case study proofs the importance of the integrated approach between urban and transport planning, especially concerning the opportunity of providing public rail transports within the urban area.

The implementation of the city plan is an important step useful for check and evaluate strategic choices, trough land use detailed plan design, integrated with transport planning design choices. In this perspective transport infrastructures became new urban elements useful to shape new transit-oriented neighborhood around them.

An integrated approach in urban planning could give effective answers to mobility questions. Urban plans could promote use of public transports and redirect growth process toward a sustainable model, characterized by mixes of uses, social cohesion and high level of accessibility.

In this new perspective, plans can provide to change the role of public transport, from a simply infrastructure to a central element, useful to set up innovative development strategies, for the requalification of the station sites, maximizing the potential of public transport and accessibility of places and make the city a fair place to live in.

The presented experience is an example of the opportunity to verify transit development models, in a medium size metropolitan city growth without consider public transports, accessibility and urban quality themes.

A first important step, consist in providing the full integration between transit-oriented policies and other planning policies, both at the regional and urban scale. That allows to address planning choices toward the use of mass rapid transports in new urban areas. In this perspective the integration between urban renewal policies and mobility plans seems to be extremely important.

Encouraging high density, energy efficiency, mixed uses, in proximity of new transit stations may drastically reduce travel long distance among the city, and, as consequence, the need of private means of transport.

Many European cities have started successfully integration policies between land use and transport planning. Following the European example, Italian cities, could start immediately with integration policies in order to address the future development toward a more sustainable model.

References

Appleyard, B., Frost, A.R.: Liveability as a framework for understanding and guiding transportation and land use integration. In: Deakin, E. (ed.) Transportation, Land Use and environmental Planning, pp. 151–167. Elsevier, Amsterdam (2019)

Barbarossa, L., Pappalardo, V., Martinico, F.: Building the resilient city. Strategies and tools for the city masterplan. J. Urban Plann. Landsc. Environ. Des. 3(2), 15–24 (2018). https://doi.org/10.6092/2531-9906/6099

Calthorpe, P.: The Next American Metropolis. Princeton Architectural Press, New York (1993)

Campos Venuti, G.: Il trasporto su ferro per trasformare le città: Roma a confronto con le metropoli Europee. Urbanistica 112 (1999)

Cervero, R., Ferrel, C., Murphy, S.: Transit Oriented Development and joint development in the United States: a literature review in "TCRP Research Result Digest" n. 52, October 2002

Curtis, C.: Public transport-orientated development and network effects. In: Hickman, R., Givoni, M., Bonilla, D., Banister, D. (eds.) Handbook on transport and development, pp. 136–148. Edward Elgar Publishing, Cheltenham (2015)

Deakin, E.: Integrating transportation, land use, and environmental planning. In: Deakin, E. (ed.) Transportation, Land Use and environmental Planning, pp. 569–600. Elsevier, London (2019)

Euromobility: La mobilità sostenibile in Italia. Indagine sulle principali 50 città, Edizione 2021. https://www.euromobility.org. Accessed 16 Oct 2021

Kenworthy, J.: An international comparative perspective on fast rising motorisation and automobile dependence in developing cities. In: Dimitriou, H., Gackenheimer, R. (eds.) Transport Policy Making and Planning for Cities of the Developing World, Edward Elgar, London (2010)

Hickman, R., Banister, D.: Transport, Climate Change and the City, Routledge, London, New York (2014)

Hickman, R., Givoni, M., Bonilla, D., Banister, D.: The transport and development relationship. In: Hickman, R, Givoni, M., Bonilla, D., Banister, D. (eds.) Handbook on Transport and Development, pp. 3–18. Edward Elgar Publishing, Cheltenham (2015)

ISTAT – Istituto Nazionale di Statistica: Censimento della Popolazione e delle Abitazioni 2011. http://www.demo.istat.it/. Accessed 15 Oct 2021

Inveroth, S.P., Johansson, S., Brandt, N.: The potential of the infrastructural system of Hammarby Sjöstad in Stockholm, Sweden. Energy Policy **59**, 716–726 (2013). https://doi.org/10.1016/j.enpol.2013.04.027

Knowles, R.D.: Transit oriented development in Copenhagen, Denmark: from the finger plan to Ørestad. J. Transp. Geogr. **22**, 251–261 (2012)

La Greca, P., Martinico, F.: Shaping the sustainable urban mobility. In: Papa, R., Fistola, R., Gargiulo, C. (eds.) The Catania case study in smart planning, sustainability and mobility in the age of change, pp. 359–374. Springer, Cham (2018)

La Greca, P., Barbarossa, L., Ignaccolo, M., Inturri, G., Martinico, F.: The density dilemma. a proposal for introducing smart growth principles in a sprawling settlement within Catania Metropolitan Area. Cities **28**, 527–535 (2011)

Marshall, S., Banister, D.: Land Use and Transport: European Research Towards Integrated Policies. Elsevier, London (2007)

Secchi, B.: A new urban question. In: Symposium Trilogy of Swiss Spatial Sciences Framework, ETH, Zurich (2010)

Identifying Spatial Opportunities for Nature-Based Solutions Planning in Cities: A Case Study in the Area of Valletta, Malta

Davide Longato[1], Chiara Cortinovis[2], Mario Balzan[3], and Davide Geneletti[1(✉)]

[1] Department of Civil, Environmental and Mechanical Engineering, University of Trento, Trento, Italy
davide.geneletti@unitn.it
[2] Department of Geography, Humboldt-Universität Zu Berlin, Berlin, Germany
[3] Institute of Applied Sciences, Malta College of Arts, Science and Technology, Paola, Malta

Abstract. Nature-based solutions (NbS) in cities are actions that utilize ecosystem processes of green-blue infrastructure to safeguard or enhance the delivery of ecosystem services and contribute to address urban challenges. The identification of spatial opportunities for NbS can support the development of concrete options of NbS planning and implementation. This research presents an approach to identify spatial opportunities to implement NbS within the urban agglomeration around Valletta, Malta. Spatial opportunities for NbS are identified through spatial analysis of available open spaces and content analysis of the spatial policies promoting NbS interventions. Overall spatial opportunities cover the 14% of the case study area. They include open spaces that are potentially available for the creation of new ecosystems, built-up areas where it is possible to integrate green elements, and existing ecosystems to conserve and/or enhance. The identification and mapping of spatial opportunities can support the NbS planning and implementation on the ground, while highlighting in which city areas there is the need to integrate alternative solutions because of the lack of space.

1 Introduction

Over the past years, an increasing number of perspectives have reflected an anthropocentric view of the management of nature and natural resources, including biodiversity and the environment, focusing on the benefits that humans gain from nature [1]. Although sharing a similar root with more consolidated concepts such as ecosystem services (ES) and green and blue infrastructure, the emergence of the notion of Nature-based Solutions (NbS) denotes the recent expansion of the scope to particularly encompass the use of nature for addressing (i.e., resolving or mitigating) multiple environmental, socio-economic, and ecological challenges [2]. For this reason, it is directly relevant and enforceable to several policy areas such as land use and spatial planning [3]. NbS can be described as actions that utilize ecosystem processes of green and blue infrastructure to safeguard or enhance the delivery of ES [4]. The promotion of NbS in urban areas builds on the increasing evidence and experiences showing that natural resources can

© The Author(s), under exclusive license to Springer Nature Switzerland AG 2022
D. La Rosa and R. Privitera (Eds.): INPUT 2021, LNCE 242, pp. 104–112, 2022.
https://doi.org/10.1007/978-3-030-96985-1_12

play an important and cost-effective role in addressing the challenges of cities, such as climate mitigation and adaptation, air pollution, and human well-being.

The availability and distribution of green and blue infrastructure elements in cities are directly linked to urban planning decisions, together with the spatial distribution and vulnerability profile of population and physical assets [5]. The relationship between urban planning and NbS therefore stems from the fact that planning practices can influence the existence, spatial extent and allocation, and even the management of green and blue infrastructure, while controlling and influencing the distribution of population and physical assets that in turn create the demand for ES to address the existing challenges. To act as an effective solution, NbS must then be carefully planned and distributed to target - in space and time - the issues and challenges affecting a city, a neighbourhood, or a specific site, while providing benefits to as many beneficiaries as possible. However, the urban form of the city represents a strong limitation, with dense urban form itself, many competing uses for land, and land ownership being important factors potentially hindering NbS mainstreaming in planning [6], especially concerning NbS that require space on the ground. To increase the uptake of NbS in urban planning practices, there is a need to know where and how much space exists for their implementation. Spatial opportunities for NbS represent possible locations where proper conditions exist for their implementation on the ground [7]. The identification of spatial opportunities for NbS is a key step towards identifying, planning, and actually implementing NbS, and can support the development of concrete options of NbS [8].

This research presents an approach to identify spatial opportunities for NbS in cities based on a case study application, and it discusses possible implementation options in the different available spaces. The case study is represented by the dense urban area around Valletta, in the small island state of Malta. Besides the identification of physically available open space for NbS through spatial analysis, spatial policies adopted in the urban plans covering the study area are analysed to identify further opportunities and options for implementing NbS that cannot be identified through simple spatial analysis of open spaces (e.g., integration of NbS into the existing built-up spaces, public spaces and infrastructure, etc.).

2 Study Site

The study site includes the capital city of Malta, Valletta, and the surrounding compact urban agglomeration constituted by numerous urban localities that form a unique urbanised continuum. The case study area is defined by the boundaries of the North Harbour and Grand Harbour Local Plan areas, covering a total surface of 2363 ha. Urban land uses cover a significant proportion of the study area, with almost 80% of artificial surfaces, while agricultural and natural/seminatural areas cover respectively 7% and 12% (Fig. 1). It includes one of the major areas for tourism along the coastal belt from Sliema to Paceville, densely populated residential areas, heavy industrial uses together with maritime-related activities, and several areas with significant natural elements (i.e., valley areas) and urban greenery (i.e., urban open spaces and green areas).

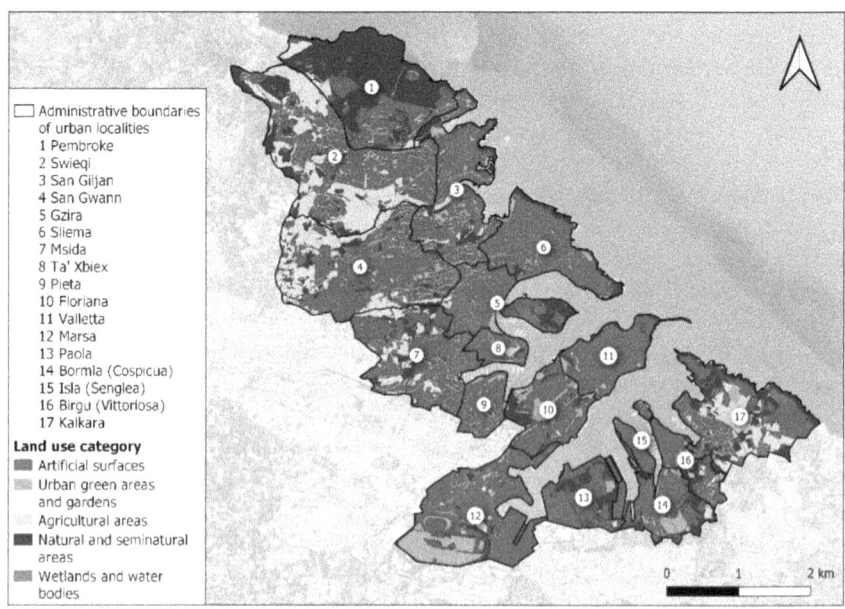

Fig. 1. Administrative boundaries of the case study area and main land uses.

3   Methodology for Identifying Spatial Opportunities for NbS

The approach used to identify spatial opportunities for NbS is based on two steps.

First, we identified open spaces (of undeveloped land) potentially suitable for the implementation of NbS on the ground. These are called physical opportunities. In order to avoid over-estimating the amount of space for effective urban green [9], as well as the land take of areas that are not intended for urban uses such as farmland and natural spaces, only open spaces located within the urban footprint were considered. The urban footprint is represented by the area located within the development boundaries and/or urban conservation areas. Development boundaries enclose the area within which it is possible to apply for building permits and, thus, where land use change to urban uses is allowed. Urban conservation areas cover already urbanised areas where special attention is paid for the historical and landscape character of the built environment. The map of open spaces with green infrastructure categories developed for the case study area in 2017 during the EnRoute project (https://oppla.eu/casestudy/19309) – further updated through photo interpretation to account for land use changes during the 2017–2020 period – was used as baseline for the identification of the physical opportunities by excluding open spaces outside the urban footprint or characterised by land cover categories that are unsuitable for land transformation (i.e., cliffs, beaches, wetlands, watercourses, garden areas, and open spaces within major government institutions). Consequently, physical opportunities mainly cover peri-urban zones potentially destined to city expansion, infill development sites, decommissioned sites, and urban open spaces that are preserved from development.

Second, we mapped the areas and sites identified by the local plans' spatial policies and regulations as target spaces for NbS interventions, namely for the conservation, enhancement, or restoration of existing, and creation of new ecosystems [10]. These are called opportunities related to planning regulations. We carried out a qualitative content analysis of the local plans to identify the spatial policies promoting NbS interventions (i.e., policies involving planning actions that explicitly include green elements, such as trees, green spaces, urban parks and playgrounds, and any kind of greenery associated with specific sites) and the related target areas and sites. They include development areas where developers are required to include green elements, green/open spaces to preserve from development, and public spaces (e.g., streets, plazas, pedestrian zones) and other sites (e.g., industrial activities, office complexes) to enhance through environmental improvements (e.g., street greenery, planting and landscaping measures for aesthetics or mitigation purposes). Once mapped, they were added to the physical opportunities to set up the final map of spatial opportunities for NbS.

4   Spatial Opportunities for NbS

A total of 332 ha of spatial opportunities for NbS were identified, corresponding to the 14% of the case study area. 207 ha are covered by physical opportunities and 188 ha by opportunities related to planning regulations, with 63 ha covered by both typologies. When overlapping, the two typologies cover open spaces designated by the spatial policies to future development schemes or to ecosystem conservation and/or improvement actions. Opportunities related to planning regulations that do not overlap the physical opportunities instead cover elements of the built environment that are not identifiable through open space analysis, such as streets and other public and private built-up sites. Table 1 shows the different land use and cover categories characterising the areas mapped as spatial opportunities for NbS.

As shown in Fig. 2, significant physical opportunities are located in the northern part of the study area, namely in the urban localities of Pembroke and Swieqi, in the western part of Msida, in Manoel Island, in Floriana and the southern part of Cospicua and Birgu. The central part with the touristic coastal belt has the lowest presence of spatial opportunities for NbS, especially in the localities of Sliema, San Julian, Gzira, San Gwann, and Pieta, together with Valletta and Senglea, the two historical and most compact cities.

The distribution of the opportunities related to planning regulations is quite scattered. Floriana, Marsa, and Cospicua show significant opportunities, contrary to the capital city of Valletta and, to a lesser extent, the towns of Senglea, Sliema, San Gwann, and Pieta. Figure 3 shows the spatial distribution of the opportunities related to planning regulations classified in four main typologies of NbS interventions based on the scope and target area of the intervention: conservation of open and green spaces, sites for urban transformation and/or development that require the integration of green elements, environmental improvement of public spaces, and environmental improvement of existing urban sites. Respectively, they cover 106 ha, 42 ha, 28 ha, and 31 ha. However, except the areas for open and green space conservation that fully correspond to the available space for NbS implementation (like the physical opportunities), the areas corresponding to the

Table 1. Land use and cover categories of the areas mapped as spatial opportunities for NbS.

Typology of spatial opportunities	Land use and cover category	Notes
Physical opportunities, including opportunities related to planning regulations that overlap them	Brownfield land, or land within urban zones which was not developed and is not used for agriculture	Typically disturbed through human action
	Abandoned agricultural areas	Land use change to urban used allowed
	Agricultural areas (arable, permanent crops)	
	Natural grassland	Land use change to urban used allowed, except for conservation areas
	Shrubland	
	Woodland	
Opportunities related to planning regulations that do not overlap physical opportunities	Residential areas	-
	Commercial areas	-
	Industrial areas	-
	Roads and associated land	-
	Other public spaces	Squares and pedestrian areas, waterfront areas

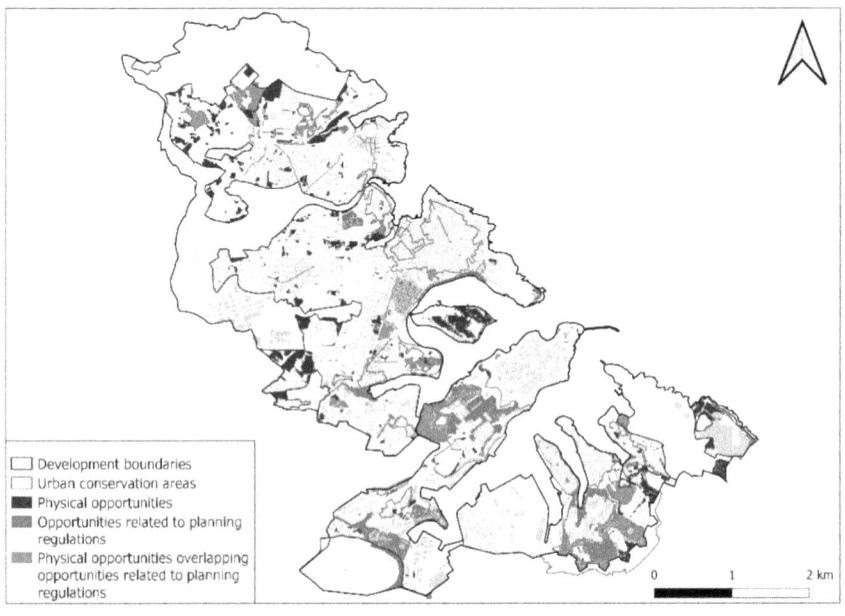

Fig. 2. Spatial distribution of the spatial opportunities for NbS.

other categories do not correspond to the available space for NbS, which are expected to be implemented only in a portion of the mapped space. For example, in the case of a street or public space identified for environmental improvement the space available for the greening intervention (e.g., street trees, public greenery) will be a portion of and not the whole street/public space area mapped.

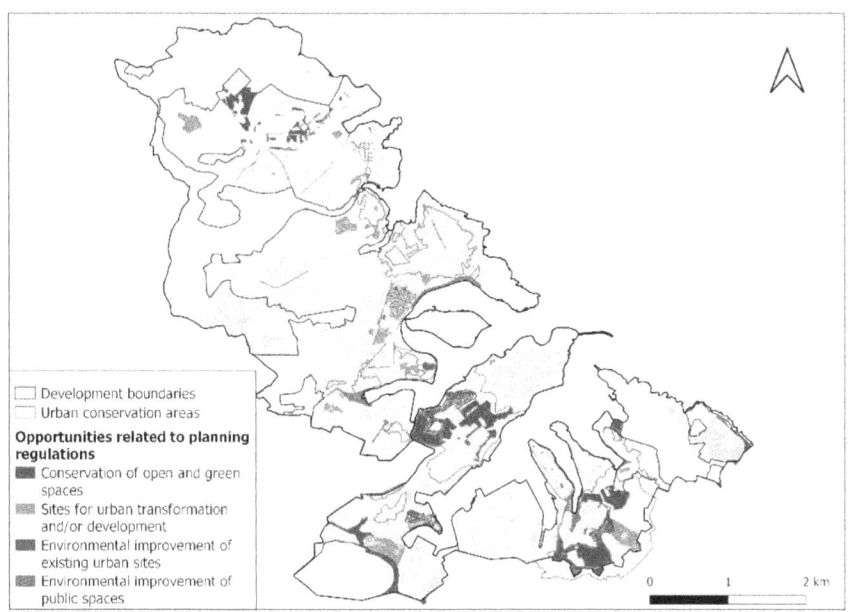

Fig. 3. Spatial distribution of the different typologies of opportunities related to planning regulations.

5   Discussion and Conclusions

Our research identified spatial opportunities for NbS on the ground in the urban agglomeration around Valletta, the capital city of Malta. We detected and mapped available open spaces through spatial analysis (called physical opportunities) and the areas and sites identified by the urban planning instruments as target spaces for NbS interventions through content analysis of the local plans' spatial policies (called opportunities related to planning regulations). Some of the areas mapped as opportunities related to planning regulations overlap the physical opportunities, namely those targeting open space areas. The others are targeted to built-up spaces (e.g., streets and other public spaces, existing built-up sites, etc.), thus allowing the identification of further potential for implementing NbS that otherwise is not possible to capture solely on the basis of the spatial analysis. The two typologies of spatial opportunities can therefore be considered complementary in providing a more comprehensive picture of the space that is potentially available for implementing NbS, which could include: the creation of new ecosystems within the available open spaces through medium- and large-scale NbS projects (e.g., urban forests and parks); the integration of small-scale NbS interventions into the built environment (e.g., trees, playgrounds, and sustainable urban drainage systems), also through restoration interventions (e.g., de-paving public spaces); the conservation and/or enhancement of existing urban ecosystems (e.g., formal and informal green spaces). Moreover, the identification of the opportunities related to planning regulations revealed possible NbS implementation options that can be promoted through specific policies and instruments and potentially scaled up within the rest of open space areas identified as physical opportunities. These include the definition of standards and requirements to apply when transforming an area (e.g., from agricultural to residential land use) with the objective to integrate NbS in the project, the definition of natural conservation zones or open space areas to be preserved from development and dedicated to greening interventions, and the promotion of best practices and criteria that include NbS in the design and improvement of public and private spaces. In particular, areas for new development and transformation projects by privates could integrate greening elements early on during the planning process with the interventions paid off by private developers, while their integration into the existing private spaces requires retrofitting interventions that can be mainly promoted through economic incentives provided by the public.

Overall, the spatial opportunities identified are not equally distributed among the study area and, for this reason, there may be significant city areas that cannot benefit from their implementation. However, a fair number of (public) open spaces that were not initially included within the spatial opportunities – because of covering areas with already established uses and functions, thus not meant for generic land use transformation and development – could offer further opportunities for NbS in such city areas that lack proper spatial opportunities. For example, existing public gardens could be re-designed not only to fulfil the recreational functions but also to accommodate, where possible, specific NbS features to address urban challenges without affecting the recreational value and accessibility. Possible solutions may include floodable areas within specific portions of urban parks to reduce stormwater runoff, increased canopy cover to provide more shadow and reduce temperature and air pollution, and increased permeable soil in highly paved garden areas to improve water infiltration and carbon storage. Another opportunity is offered by the incorporation of green elements that go beyond the mere aesthetic purpose in the street and public space greenery to mitigate specific issues at the local scale, especially those requiring ES supplied by linear green infrastructures such as noise reduction and moderation of extreme events [5]. Possible NbS that may be introduced in such spaces are vegetation buffers to shield traffic noise or linear rain gardens and bioswales to reduce stormwater runoff in streets and highly impervious public areas. However, site-specific considerations are required to assess if enough space and technical feasibility exist for introducing such elements (e.g., vegetated noise barriers require a minimum width and multi-layered vegetation to perform the noise shielding function [11]).

The mapping of spatial opportunities for NbS could offer a valuable tool that can be used as an entry point for planning NbS distribution and implementation. For example, when combined with ES mapping and assessment including both the demand and supply side, it can support decision-making to identify priority sites for NbS interventions, whether it is the conservation of existing ecosystems to secure ES provision, or the enhancement of existing and creation of new ecosystems in areas with high ES demand. In particular, the distribution of population and physical assets determines the demand for ES [12] that, together with the spatial configuration of the societal challenges and related hazards, can be used to assess and map the distribution and magnitude of the ES demanded across a city (e.g., [11]). Such knowledge is necessary to prioritize and locate the right solution, which delivers the right ES, on the right (available) place, which can be identified in the map of spatial opportunities. Finally, the identification of spatial opportunities may help to analyse what city areas may benefit from NbS implementation on the ground and where there is instead the need to design and integrate solutions that do not require space on the ground (e.g., green roofs and walls).

Funding Information. DL, MB e DG acknowledge support from the Renature project (Promoting research excellence in nature-based solutions for innovation, sustainable economic growth and human wellbeing in Malta) receiving funding from the European Union's Horizon 2020 research and innovation programme under grant agreement No. 809988. CC acknowledges support from the Alexander von Humboldt Foundation.

References

1. Nesshöver, C., et al.: The science, policy and practice of nature-based solutions: an interdisciplinary perspective. Sci. Total Environ. **579**, 1215–1227 (2017). https://doi.org/10.1016/j.scitotenv.2016.11.106
2. Escobedo, F.J., Giannico, V., Jim, C.Y., Sanesi, G., Lafortezza, R.: Urban forests, ecosystem services, green infrastructure and nature-based solutions: nexus or evolving metaphors? Urban Forest. Urban Green. **37**, 3–12 (2019). https://doi.org/10.1016/j.ufug.2018.02.011
3. Raymond, C.M., et al.: A framework for assessing and implementing the co-benefits of nature-based solutions in urban areas. Environ. Sci. Policy **77**, 15–24 (2017). https://doi.org/10.1016/j.envsci.2017.07.008
4. Albert, C., et al.: Addressing societal challenges through nature-based solutions: how can landscape planning and governance research contribute? Landsc. Urban Plan. **182**, 12–21 (2019). https://doi.org/10.1016/j.landurbplan.2018.10.003
5. Cortinovis, C., Geneletti, D.: A framework to explore the effects of urban planning decisions on regulating ecosystem services in cities. Ecosyst. Serv. **38**, 100946 (2019). https://doi.org/10.1016/j.ecoser.2019.100946
6. Johns, C.M.: Understanding barriers to green infrastructure policy and stormwater management in the City of Toronto: a shift from grey to green or policy layering and conversion? J. Environ. Planning Manage. **62**(8), 1377–1401 (2019). https://doi.org/10.1080/09640568.2018.1496072
7. Guerrero, P., Haase, D., Albert, C.: Locating spatial opportunities for nature-based solutions: a river landscape application. Water (Switzerland) **10**(12), 1–15 (2018). https://doi.org/10.3390/w10121869
8. Brillinger, M., Dehnhardt, A., Schwarze, R., Albert, C.: Exploring the uptake of nature-based measures in flood risk management: evidence from German federal states. Environ. Sci. Policy **110**, 14–23 (2020). https://doi.org/10.1016/j.envsci.2020.05.008
9. Fletcher, D.H., et al.: Using demand mapping to assess the benefits of urban green and blue space in cities from four continents. Sci. Total Environ. **785**, 147238 (2021). https://doi.org/10.1016/j.scitotenv.2021.147238

10. Cortinovis, C., Geneletti, D.: Ecosystem services in urban plans: What is there, and what is still needed for better decisions. Land Use Policy **70**, 298–312 (2018). https://doi.org/10.1016/j.landusepol.2017.10.017

11. Cortinovis, C., Geneletti, D.: A performance-based planning approach integrating supply and demand of urban ecosystem services. Landsc. Urban Plan. **201**, 103842 (2020). https://doi.org/10.1016/j.landurbplan.2020.103842

12. Langemeyer, J., Gómez-Baggethun, E., Haase, D., Scheuer, S., Elmqvist, T.: Bridging the gap between ecosystem service assessments and land-use planning through multi-criteria decision analysis (MCDA). Environ. Sci. Policy **62**, 45–56 (2016). https://doi.org/10.1016/j.envsci.2016.02.013

Integrating the Strategy for Sustainable Development in Local Planning: An Innovative Approach

Francesca Leccis[✉]

University of Cagliari, Cagliari, Italy
francescaleccis@unica.it

Abstract. On 25th September 2015 the 193 Member Countries of the United Nations signed the Agenda 2030 for Sustainable Development, which represents a global basis to build an environmentally, economically and socially sustainable world. Principles and objectives of the 2030 Agenda have been geared to the Italian context through the National Sustainable Development Strategy, approved on 22nd December 2017. Subsequently, Italian Regions formulated, or are still in the process of formulating, Regional Strategies for Sustainable Development, which are coherent with the National Strategy and with the 2030 Agenda and contribute to the achievement of their objectives and goals.

According to art.34, fourth paragraph, of the Italian legislative decree 152/2006, Local Government must implement policy instruments able to pursue the objectives of the Regional Strategy for Sustainable Development. This means that regional and local plans and programs must be consistent with the Regional Strategy for Sustainable Development and, concurrently, with the National Sustainable Development Strategy. It follows that new methodological tools are needed in order to conveniently develop public policies within the framework of both the National and Regional Strategy for Sustainable Development.

This contribution proposes an innovative procedure of urban planning that allows for the integration of Sardinia's Regional Strategy for Sustainable Development in local planning practices. The illustrated methodological approach provides a new tool, which integrates the objectives of the Regional Strategy for Sustainable Development into local plans and programs and simultaneously assesses the efficiency of these strategies in the city context. In this way, social and economic issues are taken into account alongside environmental concerns in terms of objectives, actions and indicators identified in urban and regional plans and programs, as well as in terms of participation in the planning process of economic players and social organizations together with environmental authorities.

1 Introduction

On 25th September 2015 the 193 Member Countries of the United Nations signed the Agenda 2030 for Sustainable Development, which represents a "universal call to action to end poverty, protect the planet and ensure that all people enjoy peace and prosperity by 2030" (UNDP 2021a). It seeks to tackle some of the most urgent challenges facing

© The Author(s), under exclusive license to Springer Nature Switzerland AG 2022
D. La Rosa and R. Privitera (Eds.): INPUT 2021, LNCE 242, pp. 113–122, 2022.
https://doi.org/10.1007/978-3-030-96985-1_13

the world today by balancing the economic, social and environmental dimensions of sustainability (UNDP 2021b; UNGA 2015). In particular, it identifies five critical areas of action: people, prosperity, planet, partnership and peace (Fig. 1) (UNSSC 2017).

The 2030 Agenda constitutes an opportunity to improve life for future generations ensuring that no one is left behind (UNDP 2021b), thus it covers issues that affect both planet and humanity through the definition of 17 Sustainable Development Goals and 169 targets (UNGA 2015). This means that development must balance the economic, social and environmental dimensions of sustainability (UNGA 2015). In particular, policymakers can create synergies across sectors when objectives interact positively and have to make trade-offs when they interact negatively (Nilsson et al. 2016).

Although interconnection and mutual reinforcement among SDGs are broadly recognized in the literature (UN 2005; Steffen and Smith 2013; Griggs et al. 2014; Le Blanc 2015; Lu et al. 2015; Boas et al. 2016) and several simulation models have been developed to understand synergies and trade-offs among SDGs (Sterman 2000; Barney 2002; Meadows et al. 2004; Nilsson et al. 2016; Collste et al. 2016; Weitz et al. 2017), frameworks and processes to practically implement SDGs in national and local policies are yet insufficient to effectively deal with counteractive priorities (Weymouth and Hartz-Karp 2018).

This paper illustrates an innovative methodology, developed by the DICAAR research group (University of Cagliari), to include SDGs into local plans and programs through the implementation of the strategies developed by the national government in Italy and by the regional government in Sardinia. Thanks to this methodology, local plans are defined taking into account, not only the traditional environmental issues, but also social and economic concerns identified in national and regional strategies. The next paragraph illustrates the developed methodology. Subsequently, some examples referred to the Sardinian context are reported and results are discussed. Lastly, conclusions are drawn.

2 Methodology

The methodological approach is based on the integration of the SDGs into Strategic Environmental Assessments (SEAs) of plans and programs. The relevance of the SEA in the context of pursuing the achievement of the SDGs is highlighted by both the EU (2019) and by the UNECE (2017), also in relation to their role in ensuring the understanding of conflicts and trade-offs among SDGs and in improving draft policies (ADP et al. 2019). In addition, the EU (2019) also suggests that clearer connections between the SDGs and regional and local objectives would enrich the SEA process, since they include social and economic aspects, more than environmental ones.

The integration of the selected SDGs, tailored to the Sardinian territory thanks to the definition of the Strategic Regional Objectives (SROs) of Sardinia's Regional Strategy for Sustainable Development (RSSD), is based on the Logical Framework Approach (LFA), which proved to be a valuable scheme to underline connections between project elements and context characteristics, thus emphasizing the strategy dimension (NORAD 1999) and supporting the phases of planning, design, implementation, evaluation, and follow-ups of projects and programs (Örtengren 2004; Couillard et al. 2009).

According to Örtengren (2004), the LFA is constituted by the following interrelated nine steps:

1. Analysis of the project's context
2. Stakeholder Analysis
3. Problem Analysis/Situation Analysis
4. Objectives Analysis
5. Plan of Activities
6. Resource Planning
7. Indicators/Measurements of Objectives
8. Risk Analysis and Risk Management
9. Analysis of the Assumptions

The integration of the SROs of the RSSD takes place in the fourth step, the Objectives Analysis. In particular, the implementation of the LFA led to a four-column and n-line matrix (Table 1), called the Objective tree. The four columns are: Sustainability-oriented Objectives (SO), Specific Objectives (SpO), Operational Objectives (OO) and Actions (A). The number of matrix lines is variable and depends on the objective and action counts.

Table 1. The objective tree

Sustainability-oriented objectives	Specific objectives	Operational objectives	Actions
SO.1	SpO.1	OO.1	A.1
			A.2
		OO.2	A.3
			A.4
	SpO.2	OO.3	A.5
			A.6
		OO.4	A.7
			...
...	...	...	...

More specifically, the SROs contribute to the definition of the SO reported in the first column. In fact, these are determined on the basis of both the SWOT analysis conducted in the local context, and of the SROs identified in the RSSD. However, not all the SROs are relevant to local plans and programs. Indeed, many of them can be pursued through different initiatives and tools, but cannot be addressed by regional or local plans.

For this reason, the first fundamental step to integrate the SROs into local plans and programs is the selection of the ones that are pertinent to planning practices, irrespective of the governance scale. The identification is guided by the actions that the Sardinian Regional Administration (RAS) associated with each objective, because they clarify the field of action of the objectives through the expected impact they might exert.

This methodology is implemented in the next section, where results are reported and discussed through some examples.

3   Results and Discussion

The selection of the SROs pertinent to planning practices is based on the analysis of their content and of the purposes of the related actions. The reasoning is displayed in a three-column table, where SROs, actions and evaluations are reported (Table 2). The first column lists all the 36 SROs defined in the RSSD. The second column presents the actions related to each SRO. If the action can be performed by a territorial plan, the cell is colored in green, otherwise in red. The third column showcases the output of the evaluation: when at least one action is colored in green, the SRO is considered pertinent to planning initiatives and the corresponding cell is colored in green; whereas, if none of the actions is colored in green, the SRO is considered not relevant to planning initiatives, and the corresponding cell is colored in red.

The same reasoning is applied to all the 36 SROs of the RSSD, so that 27 of them are finally excerpted from the list to be potentially included in the Objective tree of urban and regional plans and programs. This selection is then followed by the analysis of each SRO in relation to the particular context and the specific plan or program in the course of definition.

The second step of the methodology is the assessment of policy consistency and coherence which consists in analyzing the contents and objectives of Plans and Programs in force in the Sardinian Region at the European, National and Regional level (RAS 2020). For the scope of this study, the National Sustainable Development Strategy (NSDS) is analyzed and its objectives are compared and evaluated in relation to SROs and finally paired with the most appropriate SRO, on the basis of their content and aim. Table 3 displays an example of these connections, related to the green-colored SRO in Table 2. Since the NSDS provides the association of its objectives with the 2030 Strategic choices, these are reported in Table 3 as well.

The third step of the methodology is the acquisition of the association of the NSDS objectives with the Local Planning Strategic Actions (LPSAs), and related indicators, agreed by four Italian Regions (Marche, Umbria, Liguria and Piedmont) as part of a collaboration project. Table 4 shows two examples of LPSAs, and their related indicators, linked to the first NSDS objectives reported in Table 3. The other three NSDS objectives in Table 3 are not linked to any LPSAs.

Table 2. Selection of strategic regional objectives.

SROs	Actions	Evaluation
Conservation and enhancement of cultural and natural attractions of the territory.	Enhancement of cultural and natural heritage offering through diversified and flexible fruition.	
	Improvement of physical, sensory and virtual accessibility to cultural and natural attractions through performing arts and the removal of architectural barriers.	
	Enhancement and integration of museum exhibitions and natural site visits through Partner Museums Network Initiatives.	
	Development of slow-mobility systems.	
	Strengthening of certification/qualification systems of museums and visitor centers of natural sites and connection to reward systems and structures.	
	Training of displaced workers to move to jobs in archeological and natural sites.	
Improvement of accessibility to employment and promotion of self-employment.	Stimulation of job creation.	
	Facilitation of access to the labor market.	
	Quality improvement of services offered to individual citizens and businesses.	

The fourth step is the association of the NSDS objectives with the suited Indicators defined by the Italian National Institute of Statistics (ISTAT). Table 5 presents ISTAT indicators related to the NSDS objective reported in Table 4.

Lastly, descriptions of indicators, reported in Table 6, clarify which changes have to be measured over time and guide data gathering and analysis.

At this point, once the selected 27 SROs have been associated with NSDS objectives, 2030 Strategic choices, LPSAs and related Indicators, it is possible to compare them to the objectives deriving from the SWOT analysis and to consequently define the SO to be included in the Objective tree. The objectives can be either transposed in the Objective tree according to the original formulation they had in the RSSD or in the SWOT analysis or they can rather be rephrased by combining the two.

Table 3. Connections between SRO, NSDS objectives and 2030 strategic choices.

SRO	NSDS objectives	2030 strategic choices
Conservation and enhancement of cultural and natural attractions of the territory	Ensure the development of potential and the sustainable management of territories, landscapes and cultural heritage	Create resilient communities and territories, protect landscapes and cultural heritage
	Provide inclusive education for the most disadvantaged, marginalized and discriminated social groups Promote social and employment integration of young people and unemployed adults by offering professional training	Education
	Launch and set up pilot initiatives oriented towards a greater understanding of landscape and natural heritage, targeted to different groups among the general public, to be properly monitored and assessed in time	Preservation of cultural and natural heritage
	Contribute to economic diversification - particularly in rural, mountain and inner areas- to income generation and employment, to sustainable tourism promotion, to urban development and environmental protection, to cultural tourism industry support, to valorization of local handicraft and traditional crafts recovery	Preservation of cultural and natural heritage

Table 4. Connections between NSDS objectives, LPSA and related Indicators.

NSDS objectives	LPSAs	Indicators
Ensure the development of potential and the sustainable management of territories, landscapes and cultural heritage	Conservation and enhancement actions on existing historic heritage Actions oriented to landscape maintenance and protection	Number of Listed Buildings and Conservation Areas Archeological areas (m^2) Buildings located in the historic city center that are renovated or refurbished Actions oriented to landscape maintenance and protection Listed Buildings and Conservation Areas in good conservation status Archeological areas in good conservation status

Table 5. Connections between NSDS objectives and ISTAT Indicators.

NSDS objectives	Indicators
Ensure the development of potential and the sustainable management of territories, landscapes and cultural heritage	Illegal building Soil consumption and soil sealing Annual growth rate of organic farming

Table 6. Indicator descriptions.

Indicators	Indicator descriptions
Illegal building	Number of illegal buildings erected in a year every 100 authorized buildings
Soil consumption and soil sealing	Square meters of annually sealed soil per inhabitant
Annual growth rate of organic farming	Ratio between land used for organic farming in two consecutive years, %

4   Conclusion

Following the holistic approach to development suggested by the 2030 Agenda, this paper proposes an innovative procedure of urban planning to include economic, social and environmental dimensions of sustainability in planning practices by integrating the RSSD in urban and regional planning practices.

The developed methodology allows for the integration of SDGs into SEAs of plans and programs, thanks to an innovative approach, which enables professionals and practitioners to effectively deal with counteractive priorities. In particular, the SDGs are tailored to the Sardinian context through the definition of the RSSD SROs, whose integration is based on the LFA and it is structured into five steps:

1. Selection of SRO pertinent to planning initiatives.
2. Assessment of policy consistency and coherence and subsequent connections between SROs, NSDS objectives and 2030 Strategic choices.
3. Connections between NSDS objectives, LPSAs and related Indicators.
4. Connections between NSDS objectives and ISTAT Indicators.
5. Description of ISTAT Indicators.

At the end of this process the identified objectives are compared with those previously defined on the basis of the SWOT analysis in order to articulate the SO to be included in the Objective tree. This comparison can lead to the exact reproduction of both the objectives according to their original formulation in the Objective tree, or to a rephrasing of the two in order to combine them in a single objective to be included in the Objective tree. SO, defined by integrating SROs and objectives deriving from the SWOT analysis, will be further classified in specific objectives, operational objectives and actions, as illustrated in Table 1.

In this way, the LF includes objectives, actions and indicators that take into account social and economic concerns, in addition to environmental issues. Hence the purpose of the 2030 Agenda of building an environmentally, economically and socially sustainable world is pursued through the coherent strategies defined at the national and regional level and through the effective actions provided by local and sectoral plans, inspired by the SROs.

The illustrated methodology will be applied to define two territorial plans in the Italian Autonomous Region of Sardinia, the Local plan for the Municipality of Cagliari and the Park plan for the Regional park of Tepilora.

Funding. This research is implemented within the research project SOSLabs. Laboratori di ricerca-azione per la Sostenibilità urbana" [SOSLabs. Research-action laboratories for urban sustainability], financed by the Ministry of the Environment and of the Protection of the Territory and the Sea of the Italian Government within the "Bando per la promozione di progetti di ricercaa supporto dell'attuazione della Strategia Nazionale per lo Sviluppo Sostenibile - Bando Snsvs 2" ["Public selection for the promotion of research projects focusing on the implementation of the National Strategy for sustainable development – Public selection Snsvs 2"].

References

ADP (Asian Development Bank), UN (United Nations) and UNEP (United Nations Environment Programme): Strengthening the Environmental Dimensions of the Sustainable Development Goals in Asia and the Pacific. Tool compendium (2019). https://doi.org/10.22617/TIM190 002-2

Raszkowski, A., Bartniczak, B.: On the road to sustainability implementation of the 2030 agenda sustainable development goals (SDG) in Poland. Sustainability **11,** 366 (2019). https://doi.org/10.3390/su11020366

Barney, G.O.: The global 2000 report to the president and the threshold 21 model: influences of dana meadows and system dynamics. Syst. Dyn. Rev. **18**, 123–136 (2002). https://doi.org/10.1002/sdr.233

Boas, I., Biermann, F., Kanie, N.: Cross-sectoral strategies in global sustainability governance: towards a nexus approach. Int. Environ. Agreements Polit Law Econ. **16**(3), 449–464 (2016). https://doi.org/10.1007/s10784-016-9321-1

Collste, D., Pedercini, M., Cornell, S.E.: Policy coherence to achieve the SDGs: using integrated simulation models to assess effective policies. Sustain. Sci. **12**(6), 921–931 (2017). https://doi.org/10.1007/s11625-017-0457-x

Couillard, J., Garon, S., Riznic, J.: The logical framework approach-millennium. Proj. Manag. J. **40**(4), 31–44 (2009)

European Parliament: Common Provision Regulation. New rules for cohesion policy for 2021–2027 (2018). https://www.europarl.europa.eu/RegData/etudes/BRIE/2018/625152/EPRS_BRI(2018)625152_EN.pdf. Accessed 13 May 2021

EU (European Union): Study to support the REFIT evaluation of directive 2001/42/EC on the assessment of the effects of certain plans and programmes on the environment (SEA Directive) (2019). https://ec.europa.eu/environment/eia/pdf/REFIT%20Study.pdf Accessed 13 May 2021

Global Taskforce of Local and Regional Governments: Statement of the local and regional governments constituency, adopted by the regional and local authority Forum in the context of the High Level Political Forum, UN Headquarters, New Yorkon,16 July 2018

Griggs, D., et al. An integrated framework for sustainable development goals. Ecol. Soc. **19**(4), 49. https://doi.org/10.5751/ES-07082-190449

Hacking, T.: The SDGs and the sustainability assessment of private-sector projects: theoretical conceptualisation and comparison with current practice using the case study of the Asian Development Bank. Impact Assess. Project Appraisal **37**(1), 2–16 (2019). https://doi.org/10.1080/14615517.2018.1477469

Le Blanc, D.: Towards integration at last? The sustainable development goals as a network of targets. Sustain. Dev. **23**, 176–187 (2015). https://doi.org/10.1002/sd.1582

Lu, Y., Nakicenovic, N., Visbeck, M., Stevance, A.S.: Policy: five priorities for the UN sustainable development goals. Nature **520**, 432–433 (2015). https://doi.org/10.1038/520432a

MATTM Ministero dell'Ambiente e della Tutela del Territorio e del Mare: Strategia Nazionale per lo Sviluppo Sostenibile (2017). https://www.minambiente.it/sites/default/files/archivio_immagini/Galletti/Comunicati/snsvs_ottobre2017.pdf. Accessed 06 May 2021

MiTE (Ministry of Ecological Transition): Strategia Nazionale per lo Sviluppo Sostenibile. https://www.minambiente.it/pagina/strategia-nazionale-lo-sviluppo-sostenibile. Accessed 06 May 2021

Meadows, D.H., Randers, J., Meadows, D.L.: Limits to Growth: the 30-year Update, 3rd edn. Chelsea Green Publishing, Vermont (2004)

Nilsson, M., Griggs, D., Visbeck, M.: Policy: map the interactions between sustainable development goals. Nature **534**, 320–322 (2016). https://doi.org/10.1038/534320a

NORAD (Norwegian Agency for Development Cooperation): The Logical Framework Approach (LFA): Handbook for Objectives-Oriented Planning, 4th edn. NORAD, Oslo (1999)

Örtengren, K.: A summary of the theory behind the LFA method. The Logical Framework Approach. Sida (2004). https://resourcecentre.savethechildren.net/node/2033/pdf/2033.pdf, Accessed 14 May 2021

RAS (Regione Autonoma della Sardegna). Allegato alla Delib.G.R. n. 64/46 del 18.12.2020 (2020)

RAS (Regione Autonoma della Sardegna). Costruiamo insieme Sardegna 2030. Report di Posizionamento. SVASI, Cagliari (2021a)

RAS (Regione Autonoma della Sardegna). I 5 Temi Strategici (2021b). http://www.regione.sardegna.it/j/v/2568?s=419901&v=2&c=94636&t=7. Accessed 13 May 2021

RAS (Regione Autonoma della Sardegna): Partecipa al Forum Regionale per lo Sviluppo Sostenibile. https://www.regione.sardegna.it/j/v/2847?s=1&v=9&c=94637&na=1&n=4&nodesc=1&ph=1&disp=2. Accessed 13 May 2021

Steffen, W., Stafford Smith, M.: Planetary boundaries, equity and global sustainability: why wealthy countries could benefit from more equity. Curr. Opin. Environ. Sustain. **5**(3–4), 403–408 (2013). https://doi.org/10.1016/j.cosust.2013.04.007

Sterman, J.D.: Business Dynamics: Systems Thinking And Modeling for a Complex World. Irwin/McGraw-Hill, Boston (2000)

UN (United Nations). 2005 world summit outcome [adoption resolution, 60th session]. United Nations, New York (2005). http://www.who.int/hiv/universalaccess2010/worldsummit.pdf, Accessed 3 May 2021

UNDP (United Nations Development Programme). What are the Sustainable Development Goals? (2021a). https://www.undp.org/content/undp/en/home/sustainable-development-goals.html#:~:text=The%20Sustainable%20Development%20Goals%20. Accessed 30 Apr 2021

UNDP (United Nations Development Programme). Background on the goals. https://www.undp.org/content/undp/en/home/sustainable-development-goals/background/, Accessed 30 Apr 2021

UNECE (United Nations Economic Commission for Europe). Role of Environmental Assessments emphazised as key to SDG and climate change action at Minsk meeting. Press release, 20 June 2017

http://www.unece.org/info/media/presscurrent-press-h/environment/2017/role-of-environment-assessments-emphasized-as-key-to-sdg-and-climate-change-action-at-minsk-meeting/doc.html. Accessed 13 May 2021

UNGA (General Assembly of the United Nation). Transforming Our World: the 2030 Agenda for Sustainable Development. United Nations, New York (2015)

UNSSC (United Nation System Staff College). The 2030 Agenda for Sustainable Development. UNSSC Knowledge Centre for Sustainable Development, Bonn (2017)

Vaghi, S.: Attuare L'agenda 2030 a Scala Regionale: Esperienze e Ipotesi di Lavoro. In: XXXIX Conferenza Italiana di Scienze Regionali, Bolzano 17–19 settembre 2018. http://www.poliedra.polimi.it/wp-content/uploads/Articolo-AISRE-Vaghi.pdf

Weitz, N., Carlsen, H., Nilsson, M., Skånberg, K.: Towards systemic and contextual priority setting for implementing the 2030 agenda. Sustain. Sci. **13**(2), 531–548 (2017). https://doi.org/10.1007/s11625-017-0470-0

Weymouth, R., Hartz-Karp Janette, H.-K.: Principles for integrating the implementation of the sustainable development goals in cities. Urban Sci. **2**, 77 (2018). https://doi.org/10.3390/urbansci2030077

A Model to Construct Crime Risk Scenarios Supporting Urban Planning Choices

Francesca Coppola[1]([⊠]), Isidoro Fasolino[1], Michele Grimaldi[1], and Monica Sebillo[2]

[1] Department of Civil Engineering, University of Salerno, via Giovanni Paolo II,
n.132, 84084 Fisciano (SA), Italy
{fracoppola,i.fasolino,migrimaldi}@unisa.it

[2] Department of Computer Science, University of Salerno, via Giovanni Paolo II,
n.132, 84084 Fisciano (SA), Italy
msebillo@unisa.it

Abstract. The topic of urban security has a key role in creating sustainable cities and communities. Improving the personal security conditions in relation to the occurrence of predatory crimes or incivility, as well as the perception related to such events, is an important necessity in response to the Sustainable Development Goals of the 2030 Agenda and more specifically SDG 11 – Make cities and human settlements inclusive, safe, resilient and sustainable. The proposed research addresses the issue through a quantitative model of crime risk assessment. Specifically, the model is substantiated through a spatially explicit *composite crime risk index, I_{Rc}*, which allows to analyse the criticalities of the territory, highlighting their intensity and surface extension through a crime risk map. Such an index varies significantly over time, as well as varying in space, since the variables involved are extremely dynamic. In order to control this aspect, the paper proposes the construction of a parametric model in a GIS environment with a double purpose. The first one is to automate the implementation of the crime risk map construction procedure and make it replicable in any context and at any scale. The second is to control the whole procedure in order to explore the parameters in relation to which risk levels vary most significantly. The usefulness of such a model lies in the opportunity to simulate different risk scenarios to be used as additional knowledge in the ex-ante phase of urban Plan formation in order to evaluate the proposed planning choices.

Keywords: Crime risk · Parametric model · Urban planning

1 Introduction

One of the main effects and prerequisites of a city characterized by a high quality of life is the absence of fear. The presence of secure streets and public spaces is an important necessity for the well-functioning of cities and for their sustainability (Jacobs 1961; Cozens 2007; Fasolino et al. 2018). In fact, the creation of safe, inclusive, resilient and sustainable cities is one of the Sustainable Development Goals created and promoted by the UN, in particular SDG 11.

© The Author(s), under exclusive license to Springer Nature Switzerland AG 2022
D. La Rosa and R. Privitera (Eds.): INPUT 2021, LNCE 242, pp. 123–130, 2022.
https://doi.org/10.1007/978-3-030-96985-1_14

The occurrence of predatory crimes (assaults, thefts or other violent acts) or episodes of incivility negatively affects the conditions of security, real or perceived, raising fear and changing the city lifestyle (Scialdone and Giuliano 2020). Since ancient times, the need for personal security has conditioned the way cities are built. Ditches have been dug, walls erected, in an attempt to physically separate the city with its citizens from the outside world, identified as a source of danger and threats to urban life. The difference from the past is that in the modern and postmodern city, danger is no longer outside. Security threats are inside the city (Paoloucci 2003). Therefore, it is necessary to focus and act on urban space to address this problem. The importance of the urban security issue requires a new perspective with respect to the ways in which this emergency has traditionally been addressed (Fasolino et al. 2018; Nobili 2020). The objective of making the city more secure cannot be pursued exclusively with control or repressive actions or with occasional and generalized social interventions. It is necessary to address the issue of security in a multidisciplinary and multi-level perspective, following an integrated approach in which urban planning plays a fundamental role. This article explores an ongoing research study in which the topic of urban security is addressed in terms of risk. More precisely, crime risk is defined as a function of three factors: hazard (H_c), vulnerability (V_c), and exposure (E_c) and described through a spatially explicit *composite crime risk index, I_{Rc}* (Fasolino et al. 2018; Coppola et al. 2021). Such an index is a function of extremely dynamic variables that make it significantly variable with respect to both space and time. To control this aspect, the paper proposes the construction in GIS environment, of a parametric model. This one, for structure and purpose, represents a tool of great utility in the simulation of different risk scenarios to be used as additional knowledge in the ex-ante phase of urban Plan formation in order to evaluate the proposed planning choices.

2 A GIS-Based Model to Construct Crime Risk Scenarios

The parametric model presented is constructed as a tool to support quantitative crime risk assessment carried out through a spatially explicit *composite crime risk index, I_{Rc}* (Fasolino et al. 2018; Coppola et al. 2021). The latter describes the critical conditions of the area examined, highlighting their intensity and surface extent by a crime risk map structured in 5 risk classes: very low (R1), low (R2), moderate (R3), high (R4), very high (R5). The phenomenon modelled through the I_{Rc} index is characterized by a high complexity that is reflected in the construction procedure of the associated map. It is the result of the combination, on several levels, of different basic maps. More precisely, the crime risk map is constructed as a combination of three base maps (Fasolino et al. 2018): crime hazard map (H_c), crime vulnerability map (V_c), and crime exposure map (E_c). It also introduces an intermediate step of to construct the crime impact map (D_c), which is the result of combining the vulnerability and exposure maps (Fig. 1).

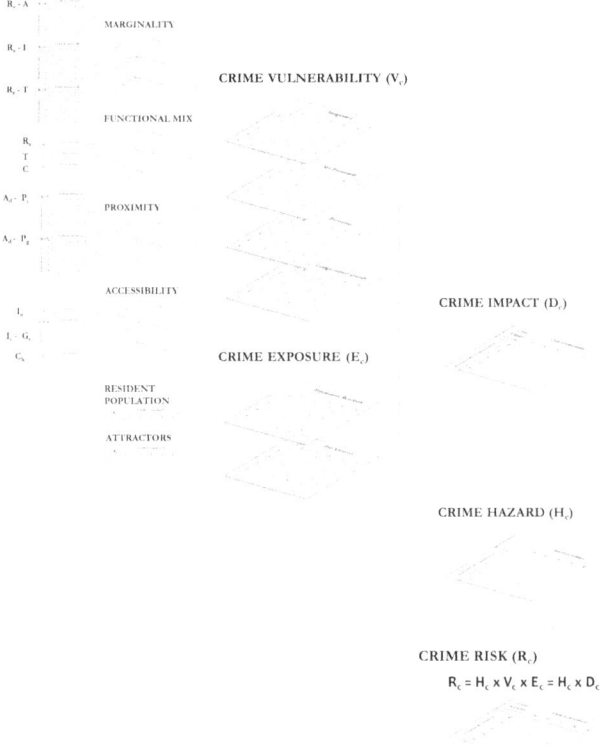

Fig. 1. Crime risk map construction scheme. Source: Authors' elaboration.

Each of the above-mentioned risk factors is in turn a function of multiple variables. Such an index as well as varying in space, varies significantly over time as the variables involved are extremely dynamic. The parametric algorithm, realized in GIS environment, has therefore, a twofold objective: to automate the implementation of the crime risk map construction procedure and make it replicable in any context and at any scale; to control the whole procedure to explore the parameters in relation to which risk levels vary most significantly. For this reason, it is a tool of great utility in the simulation of different risk scenarios, necessary to evaluate the planning choices to be proposed. The crime risk modeling through the I_{Rc} composite index is based on a matrix approach (Fig. 1) that is characterized by the use of two fundamental map algebra operations (Tomlin 1990), implemented from the base maps: *reclassification* and *overlay mapping of rasters*. The latter are necessary to combine and interpret, in a critical way, the elaboration results. For this purpose, all the maps produced are reclassified on five levels with intensity varying from very low (1) to very high (5). This choice allows to recognize the intensity degree of the phenomenon and, therefore, to govern the procedure and easily manage the combinations of several elements. Through the overlay mapping operation, a combination matrix is constructed in which each cell, defined according to the logical Boolean operator AND, is representative of a precise intensity class of the mapped phenomenon. Following this principle, and using the logic of the risk matrix (EC 2010, 2019; ECDC 2011), it is possible to interpret and reclassify the combination results by constructing five classes of crime risk: R1, R2, R3, R4, R5.

According to the objectives for which it is created, the model allows to perform spatial analysis in a short time by managing a wide variety of data. The variables involved, the parameters and the functions to be implemented in order to realize and combine the different base maps are many.

The crime risk map construction requires the definition of 16 input variables, combined on several levels, according to the risk factor described (H_c, V_c, E_c). In particular, the analysis of the hazard risk factor (H_c) is carried out by a density analysis starting from the spatial distribution of crimes occurring in the territory. The description of the vulnerability conditions to crime (V_c) is performed through density, proximity and spatial configuration analyses. For this purpose, it takes into account the spatial distribution of some particular types of land use and configurational parameters necessary for the description and study of the conditions of marginality (M), functional mix (M_f), proximity to degraded areas and buildings (P_{ro}), and accessibility (A_c). The base maps associated with these factors are the result of combining, on multiple levels, of residential (R_e), agricultural (A), industrial (I), tertiary (T), and commercial (C) land uses; of the presence and spatial distribution of degraded areas (A_d), parks (P_i), and public parking (P_g) as well as of the distribution of the global integration index (I_{in}), global choice index (C_h), connectivity index (I_c), and the presence and distribution of young residents by age group and census section (G_r) (Coppola et al. 2021). The exposure to crime conditions (E_c) are described through density analyses carried out starting from the resident population distribution, by census section (P_r) and from the presence and distribution of attractors (P_a). The procedure requires the use of 9 basic functions used for multiple purposes. Density and proximity analyses are performed by using map algebra operations of a global and focal nature through the use of the *Euclidean distance* function and the *kernel* function. Both require point-type input data that, within the model, are prepared by applying a conversion function from polygonal input data. The analysis results are then classified into criticality classes using an appropriate reclassification function. The configurational analysis is carried out with specific software, using *Linear Analysis* and, subsequently, integrated into the model. The latter manages the results and prepares them for subsequent operations through functions of buffer creation, starting from the basic topological map, and rasterization. The model foresees, moreover, the use of a dataset clipping function to restrict the analysis results to the area of interest. This can coincide with the whole municipal territory or a part of it. The overlay of the maps and the interpretation of the combination results is carried out by resorting to map algebra operations of a local nature, on which the entire modeling of the I_{Rc} index is based: *reclassification* and *overlay mapping of rasters*.

The algorithm is parameterized by defining 40 parameters, including: the cell size (l_c) and bandwidth (l_b) for implementing density analyses; a proximity criterion set through the definition of an appropriate range of distances (d_i); the buffer distance (d_b); the number (n_c) and the construction criterion of criticality classes (c_c); the maximum distance for implementing the Euclidean Distance function ($d_{e,max}$); an environment parameter to set the geographical coordinates of the examined territory to which the analyses should be extended (g_e); 10 reclassification criteria with computational purpose; 14 reclassification criteria on an interpretative basis to be applied in total to 1306 values, etc.

The model structured in this way allows the construction of the crime risk map in 89 iterations of which: 10 of converting polygonal data to point data, 1 of constructing buffers, 4 of rasterizing datasets, 8 implementations of the kernel function and 4 of the Euclidean distance function, 12 of constructing criticality classes, 13 of clipping datasets, 14 of overlay mapping and 24 of reclassification.

3   Application of the Model

The model was applied to the case study of Milan as it is an integration of a previous study that led to mapping the city' risk conditions (Fasolino et al. 2018).

The parametric algorithm was reconstructed as a Model builder through arcGIS software (Gorr and Kurland 2011).

The data used for the application were extracted from a dataset specifically constructed for the case study (Fasolino et al. 2018) by using mainly official sources such as the Milan Municipality's Geoportal, the Lombardy Region's Geoportal and Istat, with an update to 2012. In the absence of data from official sources, the spatial distribution of crimes was extracted from crime maps available online and related to the years 2010–2011.

The application was carried out using the scenarios technique, in order to explore the parameters with respect to which the I_{Rc} index varies most significantly. Starting from scenario 0, in which the crime risk map was constructed by setting the parameters according to the case study characteristics, 9 scenarios were simulated. These last ones were realized changing, from time to time, a parameter of the model, at parity of all the others. The attention was focused, in first instance, on the following parameters: cell size (l_c), bandwidth (l_b), proximity criterion (d_i, with i = 1, ..., 4), construction criterion of criticality classes (c_c), buffer distance (d_b). Some of the results obtained are shown in Fig. 2.

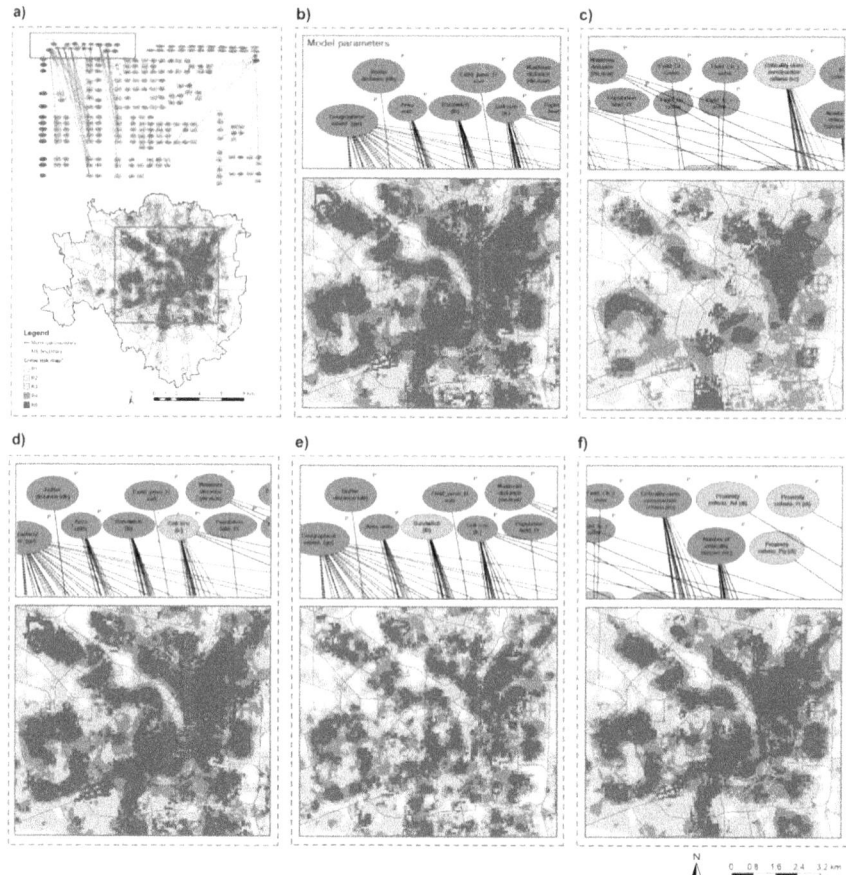

Fig. 2. a) Scheme of the parametric model created in GIS environment and crime risk map constructed for the municipality of Milan, scenario 0; b) Detail of the crime risk map, scenario 0: l_c = 20, l_b = 500 m, d_1 = 100 m, d_2 = 300 m, d_3 = 500 m, d_4 = 1000 m, c_c = Natural breaks-N_b; c) Detail of the crime risk map, scenario 1: c_c = Equal interval-E_i; d) Detail of the crime risk map, scenario 2: l_c = 1; e) Detail of the crime risk map, scenario 4: l_b = 300 m; f) Detail of the crime risk map, scenario 6: d_1 = 50 m, d_2 = 100 m, d_3 = 150 m, d_4 = 200 m. Source: Authors' elaboration.

4   Results Discussion and Conclusions

The application to the case study of Milan, even if restricted to a limited number of parameters, has identified the different influence of these ones on the crime risk map. The analysis has shown, in fact, that the bandwidth (l_b) and the criterion for the construction of criticality classes (c_c) (Fig. 2c, e) have a greater impact on the final result than the cell size (l_c) (Fig. 2d) and the buffer distance (d_b). Their variation involves a significant change in the presence and surface extent of crime risk levels, especially medium, high and very high (R3, R4, R5). In particular, the bandwidth variation can influence the readability and the detail level of the map carrying, if set on high values, to the predominance of the

R5 risk class in comparison to all the others. This situation already occurs for l_b values equal to 800m (scenario 5) and is distant from reality. Among the examined parameters, also the criterion of proximity (d_i) influences with a certain relevance the I_{Rc} index, mainly affecting the crime risk levels of the municipality peripheral areas (Fig. 2f).

Additional risk scenarios could be constructed by varying, for example, the spatial distribution of crimes using as input the outcome of a predictive analysis. Again, simulating refunctionalisation and regeneration interventions by changing the variables associated with the different land use categories and modifying the number and surface extent of degraded areas.

The results obtained outline the need for in-depth study of these aspects. In this sense, among the research development prospects are the implementation of both the sensitivity analysis on the I_{Rc} index and a statistical analysis, aimed at identifying the significance of each parameter and at model calibration.

The study presented here deepens an ongoing research but also represents an opportunity to emphasize the importance of crime risk identification, analysis and assessment for improving urban security conditions. The knowledge of the risk conditions that characterize a specific context is of great importance for the definition and design of appropriate intervention strategies and for results monitoring. The risk linked to the occurrence of criminal events is defined within the model as a convolution of three risk factors (H_c, V_c, E_c) (Fasolino et al. 2018). Its criticality classes can, therefore, vary in space and time according to the changes of each of them. The latter can be conditioned by several factors, such as social, cultural, economic aspects, but also as a consequence of planning choices.

City planning, design and construction criteria can reduce crime risk or increase it if not properly defined and assessed. To reduce this, urban security must be included from the earliest stages of planning, addressing its aspects in all urban planning instruments, whether structural plans, operational plans, sectoral plans, urban regeneration programmes, major equipment projects, etc. The action, carried out following a multi-scalar approach, makes it possible to act on all aspects contributing to the creation of secure urban spaces. Through the appropriate location of infrastructures, public spaces and services, playgrounds, activities, by defining the uses and forms of urban space (structure of space, green areas, car parks, roads, fences, trees, paths, lighting, etc.) it is possible to influence the city' s vitality and time of use, the environments perception and stimulate the citizens' sense of belonging to the territory (Fasolino et al. 2018). These elements are essential to reduce crime risk vulnerability and to increase real and perceived security. Therefore, the knowledge of the current risk conditions and of their variation caused by foreseen or proposed changes in the physical-functional structure of the territory allows to orient the urban project towards the mitigation of crime risk, instead of unconsciously contributing to its increase.

According to this, it emerges the usefulness of the constructed model whose strong point is the possibility to simulate, for any context and at any scale, different crime risk scenarios. These last ones, incorporated as knowledge elaborations in the ex-ante phase to the urban Plan formation, would allow to evaluate the proposed planning choices and to create a suitable framework of rules aimed to the city security increase.

References

Coppola, F., Grimaldi, M., Fasolino, I.: A configurational approach for measuring the accessibility of place as an analysis tool for crime risk vulnerability. In: La Rosa, D., Privitera, R. (eds.) Innovation in urban and regional planning. INPUT 2021, LNCE, vol. 146, pp. 501–509 Springer, Cham (2021). https://doi.org/10.1007/978-3-030-68824-0_54

Cozens, P.: Planning, crime and urban sustainability. WIT Trans. Ecol. Environ. **102**, 187–196 (2007). https://doi.org/10.2495/SDP070181

EC-European Commission: Commission Staff Working Paper Risk Assessment and Mapping Guidelines for Disaster Management, 21 December 2010, SEC (2010), 1626 Final. European Commission, Brussels (2010)

EC-European Commission: Recommendations for National Risk Assessment for Disaster Risk Management in EU. European Commission (2019). https://doi.org/10.2760/147842

ECDC-European Centre for Disease Prevention and Control: Operational guidance on rapid risk assessment methodology. European Centre for Disease Prevention and Control, Stockholm (2011). https://doi.org/10.2900/57509

Fasolino, I., Coppola, F., Grimaldi, M.: La sicurezza urbana degli insediamenti. Azioni e tecniche per il piano urbanistico. FrancoAngeli, Milano (2018)

Fasolino, I., Coppola, F., Grimaldi, M.: A model for urban planning control of the settlement efficiency. A case study. ASUR-Archivio di Studi Urbani e Regionali, LI **127**(suppl), 181–210 (2020). https://doi.org/10.3280/asur2020-127-s1010

Gorr, W.L., Kurland, K.S.: GIS Tutorial for Crime Analysis. Esri Press, Redlands (2011)

Jacobs, J.: The Death and Life of Great American Cities. Random House, New York (1961)

Nobili, G.G.: Le politiche di sicurezza urbana in Italia: lo stato dell'arte e i nodi irrisolti. Sinappsi X **2**, 120–137 (2020). https://doi.org/10.1485/2532-8549-202002-9.

Paolucci, G.: Il mercato della paura. In: Amendola, G. (ed.) Il governo della città sicura. Politiche, esperienze, luoghi comuni. Liguori, Napoli (2003)

Scialdone, A., Giuliano, G.A.: Percezioni di insicurezza e bisogno di protezione. Un'esplorazione della domanda sociale. Sinappsi X **2**, 43–56 (2020). https://doi.org/10.1485/2532-8549-202002-4/

Tomlin, D.C.: GIS and Cartographic Modeling. Prentice Hall, Upper Saddle River (1990).

Clustering Social Vulnerability: An Application Model

Eliana Fischer[✉]

Department of Physics and Astronomy, University of Catania, Catania, Italy
`eliana.fischer@unict.it`

Abstract. The growing vulnerability of urban areas to dangerous events is due to numerous factors: the unpredictability of some significant climate changes, the complex dynamics of urban growth, and urban policies' efficiency in providing adequate responses to the need to reduce risks.

Scientific literature and international documents in sustainable development and disaster risk reduction emphasize the importance of disaster risk management instead of disaster management alone. One of the aspects that inform the Sendai Framework for Disaster Risk Reduction 2015–2030 is implementating the experience gained through national and international strategies in plans for disaster risk reduction. The intent is to raise public and institutional awareness, trigger political commitment and involve stakeholders at different levels. What emerged from the previous decade (2005–2015) is that the consumption of goods and people increased faster than vulnerability decreased.

This work describes an applied methodology for assessing vulnerability in an urban area, through the quantification of an index. The analysis led to mapping territories with disadvantaged characteristics from the point of view of social vulnerability, using R-project and GIS software. The quantitative analysis underlines the dimensions that influence the vulnerability, highlighting evident clustering in specific geographic areas. The research represents an analytical approach to evaluate the priority interventions in the regulatory instruments (plans and programs) in line with the tendency to include prevention among the planning criteria for environmental protection against risks and ensuring the conditions for sustainable development.

Keywords: PCA · Cluster and outlier analysis · GIS · Social vulnerability index

1 Introduction

The disasters that affect urban and territorial systems show that risk and intensity are not the only factors that determine damage. The peril to which the cities are susceptible is the consequence of a complex interaction between potential physical hazards (floods, earthquakes, drought) combined with the progression of climate change with the vulnerability of social, infrastructural, economic and governmental systems (Oliver Smith 2004). Some resercher stress that the effects of a crisis are already present in the affected social system, "which manifests a specific vulnerability share for each emergency" (Ligi

© The Author(s), under exclusive license to Springer Nature Switzerland AG 2022
D. La Rosa and R. Privitera (Eds.): INPUT 2021, LNCE 242, pp. 131–140, 2022.
https://doi.org/10.1007/978-3-030-96985-1_15

2009). Therefore, the different answers to the dangers are to be found in the vulnerability of the affected social system itself (Spielman et al. 2020). The importance of the socio-economic aspects of the vulnerability had already emerged in 1994 at the first World Conference on the reduction of natural disasters organized by the International Decade for Natural Disaster Reduction. On this occasion, for the first time, the urgency of adopting strategies necessary to reduce vulnerability was affirmed. The Hyogo Framework for Action and the Sendai Framework for Action, 2005 and 2015 respectively, underline a systematic Community-based and People-Centered approach to reduce vulnerability and improve risk prevention. Cutter et al. (2003) propose a quantitative approach to social vulnerability selecting social factors that influence the susceptibility of a community towards a dangerous event and the ability of the community itself and local governments to manage the case. The social vulnerability is understood as a quantity directly related to a territory's risk and inversely with its resilience. It represents the phenomenon in its anthropic sphere concerning territory exposure to natural disasters (climatic or seismic events).

In Italy, the National Statistical Agency (Istat) adopted the concept of social and material vulnerability (IVSM-Social and material vulnerability index). Vulnerability is defined as a "ability to self-determination of subjects [...] permanently threatened by an unstable insertion within the main systems of social integration and distribution of resources" (Ranci 2002).

The points of contact between these two indicators are considerable, although the theoretical bases of the IVSM are different and disconnect the index from its link with the sources of potential risk to which contemporary societies are subjected. A recent study of Paleari (2018) highlights as 44% of the Italian territory and 36% of its municipalities are located in the seismic area 1 and 2 (the most dangerous between the 4 categories established by national legislation). Moreover, 90% of the municipalities of five regions (Marche, Molise, Basilicata, Calabria and Sicily) are in the seismic area 1 and 2. It is reductive to speak of social vulnerability without referring to the size of the risk, in a territory so widely exposed to dangers like Italy. The influence of the social factors on risk management is decisive in research on the vulnerability of urban systems and on identifying the risk prevention and mitigation strategies in territorial and urban planning as suggested by the People-Centred route expressed by the Sendai Framework for Action.

This work describes a methodology applied for the evaluation of social vulnerability to the urban scale, that ends up with the proposal of a vulnerability index. Our analysis leads to the mapping of the municipal territory with critical social characteristics by using R-Project and GIS software. The proposed model identifies the geographical areas where the main clusterization of the phenomenon is concentrated. Thanks to the adoption of the census tracts as a minimum reference unit for the analysis, this study has produced a detailed quantification of the size of vulnerability. The results are deemed useful for the definition of risk reduction policies at different scales.

2 Materials and Methods

2.1 Social Vulnerability

The calculation of the synthetic index of social vulnerability includes the following steps: i) the definition of the phenomenon object of study; ii) the selection of elementary indicators; iii) the standardization of the indicators; iv) the aggregation of the indicators in one or more synthetic indices (OECD 2008).

The analysis of literature on social vulnerability, the starting point of this work, suggested the adoption of a representation scheme of the vulnerability concerning the following material and social dimensions:

– *Family with reference to the ageing of the population and the youngest population.* Population ageing is a predominantly process of the society "vulnerabilization": both from the point of view of impaired physical mobility of the elderly (compared to the phase following a calamitous event) compared to social marginalization. Children are the other demographic group more at risk in the event of a disaster. These factors become determinants as risk drivers.
– *Level of education,* an indicator used both to evaluate human capital and, therefore, the capacity that each individual has to be able to act and cope with any adverse risks, both for its economic component as they are negatively related with aspects of material deprivation. Low levels of education can translate into more significant economic and social hardship (Istat 2020).
– *Participation in the labor market.*
– *Marginalization* defined as the condition of foreign resident.
– *Housing conditions*, including the date of construction of buildings, the maintenance status, the building type concerning the number of apartments, the under-utilization of housing stock, empty apartments, the constructive typology.

Many of the variables represent *proxy* values of the investigated phenomenon.

The official sources of the data used in the present study are two: Istat last National Census Data (year 2011) and the cartography provided by the official maps database of the Sicily Region of 2003–2004, in scale 1:2000. The first includes general information of the population and homes associated with the census tracts. The second contains the geo-referenced information relating to the buildings. The latter, implemented in GIS Environment, made it possible to calculate the volume of the residential buildings in each district area and identify the under-utilization of housing stock.

The indicators were selected both with positive and negative polarity for the phenomenon of social vulnerability. Data normalization has allowed preventing elementary indicators with higher variability to have a greater weight on the construction of the index, more influencing the results. The extraction technique adopted is the Principal Component Analysis (PCA), using the R-Project calculation software. The PCA is a methodology widely used in the field of synthetic indexes (OECD 2008; Cutter et al. 2003; Schmidthlein et al. 2008; Borden et al. 2007) as it reduces the dimensionality expressed by the number of indicators entered in the definition of the theoretical model.

2.2 Index Construction and Clustering

The PCA analysis uses a table consisting of V variables (columns) and n (census observations-sections) and returns a $V \times C$ matrix, where each line represents a variable (V) and each column is a component (C). The extraction operation, developed through the R-Project software, returns some components equivalent to the number of variables into the model. An α component is a weighted combination of variables, and every single cell of the output matrix represents the weighted contribution of each variable in the component, called "loading". The components are often assigned an arbitrary name that depends on the variables with the highest charge within the $1 \times V$ component carrier (Component-Wise approach) (Cutter et al. 2003). This phase represents one aspect of greater arbitrariness in the PCA, which is often susceptible to interpretation errors by the researcher. Unlike Cutter et al., this study proposes an approach centered on variables (Variable-Wise), identifying each component's net contribution through the components themselves (Spielman et al. 2020). In order to calculate the components' score of observations, the value is taken for that observation date x (census tracts), which the sum of several variables will represent, and each of the V variables is multiplied by the charges of the components of interest. This method allows the calculation of the specific score for each observation due to the main components m, taken into consideration for calculating the synthetic index and selected due to the autovalue greater than 1, according to the formula:

$$SoVi = \sum_{c=1}^{m} \theta_v x_{i,v} \tag{1}$$

in which $\theta_v = \sum_{c=1}^{m} \alpha$ represents the sum of the charges of each variable through the selected components; $x_{i,v}$ is the v_{th} variable for the i section.

The approach we used aims not to disperse the contribution of the individual variables introduced into the model and to optimize the interpretation of their weight within the synthetic index.

The use of spatial autocorrelation measures supports the spatial distribution of the Social Vulnerability Index and qualitative analysis: these allow to evaluate in which grade and how the value of the index is grouped into space. It describes the index trend to the aggregation and polarization by identifying similarity patterns and abnormal values. For this purpose, the Cluster and Outlier Analysis, in the GIS Environment is adopted as a clusterization technique. The spatial autocorrelation of a variable, i.e. the social vulnerability index, is the correlation between the x values in each spatial unit and the values of x in the surrounding spatial units. The I index of Moran (Anselin 1995), also defined Local Indicator of Spatial Association (LISA) used in this analysis, is one of the simplest and most widespread measurements of the spatial autocorrelation of a variable. Significantly positive values of I, i.e. spatial autocorrelation, indicate the trend of the x variable (the vulnerability index) to aggregating, or "polarizing", spatially, that is to distribute themselves according to a provision in which high (low) accompany high values (low) in an area in the surrounding areas. Non-significant levels (close to zero) of I, finally, indicate the absence of clustering (which is between values of the same sign or between opposite sign values), and therefore describe a distribution of x

tendentially random. It is required to measure this correlation to establish a proximity criterion that defines the set of surrounding spatial units for each unit. According to the census tracts, the proximity criterion adopted here is the one for which they are considered as "surrounding" the sections falling within a radius of 1000 m from the central section, useful to define the distance between a neighborhood and another. For the most appropriate spatial conceptualization, the relationship between spatial units is based on the criterion of "reverse distance", in which the nearest spatial units have a more significant influence than those more distant.

2.3 Case Study

This study focuses on the city of Catania, located in Southern-Italy (Sicily) (Fig. 1a). As shown in Sect. 2.1 the data used for analysis refer to the 2011 census and aggregated at the district level, with a resident population of 293,902 inhabitants. The city is divided into 6 districts. Table 1 shows the names of the districts, the number of census tracts and the number of inhabitants for each district.

Table 1. Districts of Catania, with number of census tracts and inhabitants.

Id	District	Number census tracts	Inhabitants
1	Monte Po - Nesima - San Leone - Rapisardi	255	44690
2	San Giorgio - Librino - S.G. La Rena - Zia Lisa	412	56208
3	Centro - S. Cristoforo	651	49483
4	Ognina - Picanello - Barriera - Canalicchio	530	64330
5	Borgo – Sanzio	347	40897
6	San Giovanni Galermo - Trappeto - Cibali	285	38294
	Total	2480	293902

3 Results and Discussions

The Social Vulnerability Index for the city of Catania was calculated for the 2480 sections, with the extraction methodology proposed in Sect. 2.2. The value of the index has been standardized and categorized according to the Standard Deviation in five classes (Table 2).

Figure 1b shows the Social Vulnerability Index distribution within Catania, the five ranges of the index and corresponding categories, where A is the less vulnerable and E is the most vulnerable.

Ognina is the district that presents more sections in the first three ranges of the Social Vulnerability Index (A-B-C; SoVi $\leq$ 0), having values ranging from 24% to 26% of the sections of the district for each range. The district with more sections in the last two ranges (D-E; 0 < SoVi $\leq$ 9) is Centre - San Cristoforo, with values that range from 30% to 41.50%. It is worth to note that high concentrations of sections in the first three ranges

Table 2. Categories and range of the social vulnerability index according to the standard deviation and number of sections for each range.

Category	Range	Number of census tracts
A	$-3.5 < \text{SoVi} \leq -1$	204
B	$-1 < \text{SoVi} \leq -0.5$	585
C	$-0.5 < \text{SoVi} \leq 0$	625
D	$0 < \text{SoVi} \leq 0.5$	498
E	$0.5 < \text{SoVi} \leq 9$	568

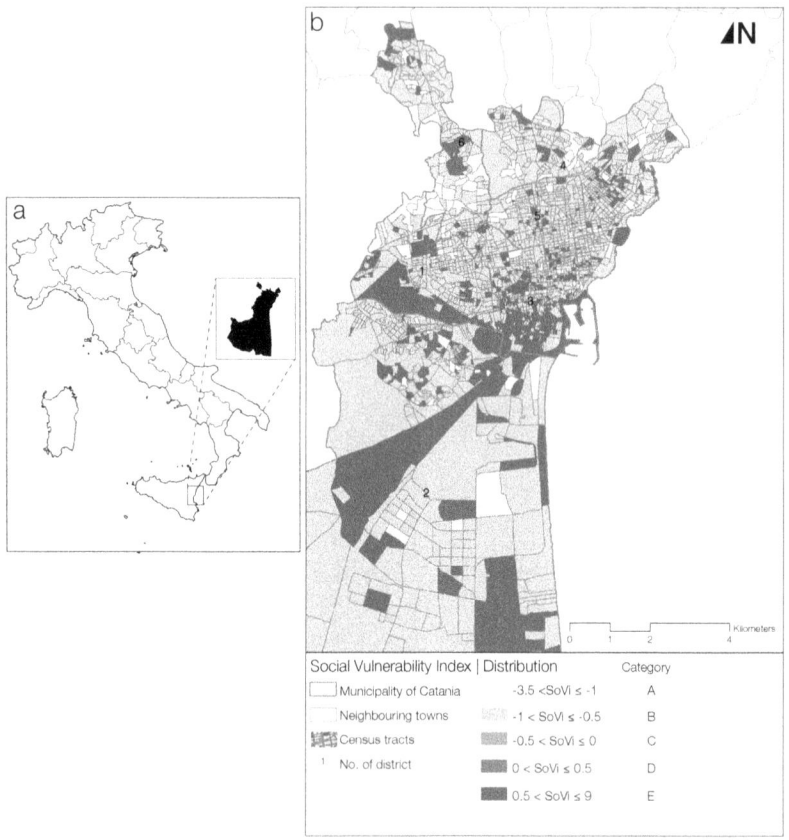

Fig. 1. a) The study area within Catania, Eastern Sicily (Italy). **b)** Spatial distribution of social vulnerability index within Catania. The map shows five ranges and corresponding categories (A–E) of the Index.

are also located in the Centre district and in the districts of Borgo and San Giorgio; high concentrations of sections in the last two ranges are located in the districts of San Giorgio and Ognina. Although it is already evident from the qualitative analysis of the categories as the high and low index values are distributed, the cluster mapping made it possible to corroborate the results of the spatial distribution of the Social Vulnerability Index. The mapping of the LISA cluster recognized for the Social Vulnerability Index allows locating groups of adjacent sections in the district having more significant statistical weight in the positive space autocorrelation of the variable, i.e. the section groups in which it occurs in a way statistically more relevant the spatial aggregation of high values of the variable with other high values ("*high-high*" groups), or low values with other low values ("*low-low*" groups). The other two types of LISA cluster, on the other hand, identify the so-called "spatial outliers", i.e. sections with high values of the index near sections with low values and viceversa ("*high-low*" type and "*low-high*" type). With the analysis of LISA groups, therefore, it can occur if and to what extent the considerations obtained from the qualitative reading of the index find statistical confirmations in the local spatial association measures. The cartographic representation of these clusters was left in the original vector form in which the census tracts are represented, and each is attributed (if statistically significant) the relative type of cluster. The mapping of the LISA Cluster for the Social Vulnerability Index highlights the characteristics of aggregation of "*high-high*" values in the areas where the latter shows the major values and the characteristics of aggregation of "*low-low*" values where the Vulnerability Index shows minor values.

To define the aggregation rate ($I_{c,m}$ [0,1]) for every district given by the report of the Census Units in "*high-high*" and "*low-low*" on the total of the sections by district, we identified an index as represented by the formula:

$$Ic = \frac{N_{c,m}}{N_{s,m}} \tag{2}$$

in which $N_{c,m}$ is the number of clusters for district in the case of the values of the variable agreement ("*high-high*" e "*low-low*") and $N_{s,m}$ is the total number of sections present in the district. Table 3 shows values of the I_c aggregation index for each district: the highest values are in the districts Centre - San Cristoforo and San Giorgio-Librino, respectively 0.51 and 0.15 ($I_{c,m}$ [0,1]). These results are in line with the socio-economic characteristics of the two districts: San Giorgio district has more favorable demographic characteristics having an ageing index of 68% (119% in the Centre), the presence of foreign residents is almost zero (6% in the Centre). In the Centre, the characteristics of the built heritage are very unfavorable, as the buildings in the impoverished state are 13%, compared to the 3% in San Giorgio, almost 100% of buildings were constructed before the normative anti-seismic, and 74% is built with unable techniques to resist the seismic event. Figure 2 shows for each district the sections that are polarized with respect to these values.

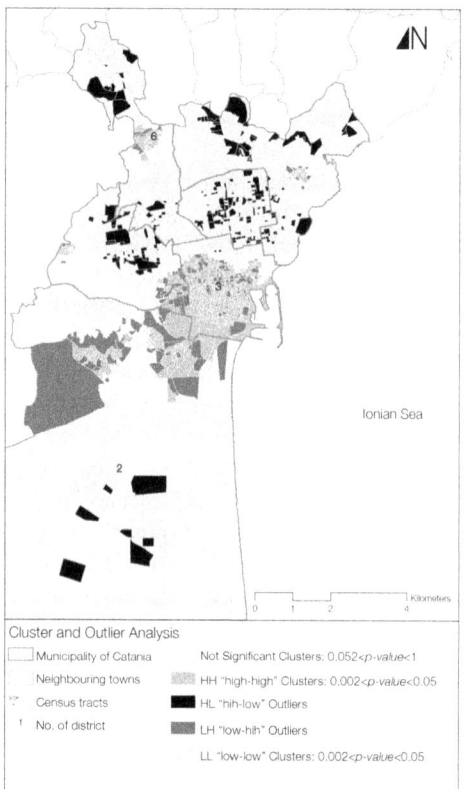

Fig. 2. Cluster and outlier analysis in the city of Catania. The map shows two Cluster groups (HH "*high-high*" – LL "*low-low*") and two Outlier groups (HL "*high-low*" – LH "*low-high*") within the six districts. They are also indicated census tracts where spatial aggregation is not-significant ($p_{value} > 0.05$).

Table 3. Aggregation index by census tract according to HH groups "*high-high*" ($I_{c,hh}$) and LL "*low-low*" ($I_{c,ll}$). The table includes the number of sections falling within the HL ("*high-low*"), LH ("*low-high*") groups, and of the sections that have a statistically not significant aggregation

	District	HH	LL	HL	LH	NS	$I_{c,hh}$	$I_{c,ll}$
255	Monte Po - Nesima - San Leone - Rapisardi	23	89	32	22	89	0.09	0.35
412	San Giorgio - Librino - S.G. La Rena - Zia Lisa	**60**	40	7	37	268	**0.15**	0.09
651	Centro-San Cristoforo	**334**	17	2	113	185	**0.51**	0.03
530	Ognina - Picanello - Barriera - Canalicchio	14	**214**	48	9	245	0.03	**0.40**
347	Borgo-Sanzio	-	**244**	83	-	20	-	**0.70**
285	San Giovanni Galermo - Trappeto - Cibali	15	88	22	5	155	0.05	0.31

4 Conclusion

The Social Vulnerability Index was built to provide a tool to support risk prevention, descriptive of some population and environmental groups' exposure levels and built to connect vulnerability factors with seismic danger.

The theme of social vulnerability has been widely analyzed mainly with its effectiveness in capturing the phenomenon of a community's reactivity compared to the manifestation of a dangerous event.

The Principal Component Analysis is one of the possible methodologies for the selection of the latent variables. One of the primary limits of the PCA is that the resulting components are complex miscellanea of the input variables, which make the naming phase highly subjective. The approach proposed in the present study aims to select all the components extracted from the PCA and not disperse the contribution of the variables, proposing a variable-wise approach instead of a component-wise one. Cluster analysis, as an analytical model, allows the identification of common aspects in the factors of vulnerability. Indeed, thanks to the identification of the polarities, it will be possible to define the mitigation programs of the vulnerability indicators. These interventions, based on the criterion of priority allocation of resources, will include targeted measures to reduce the dominant characteristics of discomfort from the socio-economic point of view, and define policies and programs on a set of groups with homogeneous characteristics of vulnerability factors. For instance, census tracts, in which the presence of the elderly is dominant, would require interventions aimed at reducing the mobility difficulties of users; a specific intervention in this direction could facilitate evacuation in the event of an earthquake. Facilitating socio-demographic diversification policies, for example through housing policies with the adoption of subsidized rents, could have the positive effects of limiting the isolation of the elderly, slowing down the process of abandonment of some neighbourhoods and maintaining the buildings.

However, various vulnerability indicators highlights the different traits of vulnerable groups, which require differentiated treatments and responses for the purpose of an effective strategy of risk mitigation and adaptation to climate change. This can be further investigated through a bottom-up participatory approach to assess and verify vulnerability by involving stakeholders. The bottom-up approach can help in understanding the causes of vulnerability and the effectiveness of some measures to reduce it, in addition to the indispensable contribution of the local community in identifying shared strategies. Therefore, for the purpose of an exhaustive interpretation of these social characteristics, once identified the most vulnerable areas, it is appropriate to carry out checks through sample surveys that allow the dual effect of confirming (or not) the hypotheses made by statistical surveys and involving the stakeholders on the other hand.

Preventive planning strategies require a transversal analysis of physical-spatial and socio-economic vulnerabilities. The two concepts have a close connection. By crossing the types (low medium high) of vulnerability (physical and social vulnerabilities), comparative studies can be developed on the basis of which setting policies aimed at satisfying the different types extracted.

The development of this part is beyond the intent of the present work.

References

Anselin, L.: Local indicators of spatial association-LISA. Geogr. Anal. **27**(2), 93–115 (1995). https://doi.org/10.1111/j.1538-4632.1995.tb00338.x

Borden, K., Schmidtlein, M.C., Emrich, C.T., Piegorsch, W.W., Cutter, S.L.: Vulnerability of U.S. cities to environmental hazards. J. Homeland Secur. Emerg. Manag. **4**(2), 1279. Article 5 (2007)

Cutter, S., Boruff, B., Shirley, W.: Social vulnerability to environmental hazards. Soc. Sci. Q. **84**(2), 242–261 (2003)

Istat: Le misure della vulnerabilità: un'applicazione a diversi ambiti territoriali (2020)

Ligi, G.: Antropolgia dei disastri, Editori Laterza (2009)

OECD: Handbook on constructing composite indicators: methodology and user guide. Joint Research Centre Europena Commission (2008)

Oliver-Smith, A.: Theorizing vulnerability in a globalized world: a political ecological perspective. In: Bankoff, G., Frerks, G. (eds.) Mapping Vulnerability: Disasters, Development and People, 1st edn. Routledge (2004). https://doi.org/10.4324/9781849771924

Paleari, S.: Natural disasters in Italy: do we invest enough in risk prevention and mitigation? Int. J. Environ. Stud. **75**(4), 673–687 (2018)

Ranci, C.: Le nuove disuguaglianze sociali in Italia. Bologna, Il Mulino (2002)

Schmidtlein, M., Deutsch, R.C., Piegorsch, W.W., Cutter, S.L.: A sensitivity analysis of the social vulnerability index risk analysis. Risk Anal. **28**(4), 1099–1114 (2008)

Spielman, S.E., et al.: Evaluating social vulnerability indicators: criteria and their application to the social vulnerability index. Nat. Hazards **100**(1), 417–436 (2020). https://doi.org/10.1007/s11069-019-03820-z

Peripheries, Rural and Cultural Landscapes

Protection from Seismic and Pandemic Risks of Fragile Suburbs. Coexistence with Risk and Distancing-Integration Scenarios

Maria Angela Bedini[✉]

Department Simau, Polytechnic University of Marche, Via Brecce Bianche, 12, 60131 Ancona, Italy
m.a.bedini@staff.univpm.it

Abstract. The research starts from the study of the different forms of marginal areas with low population density, mainly present in Central Italy, distinguished with visionary definitions: "filaments of lights", "luminous serpentines", "shining paths", "urban comets", "luminous constellations", "necklaces of lighting candles".

The research evaluates an overall strategy for the widespread residential model, supported by the resources, made available by the European Union "Recovery Plan", for large strategic projects of widespread health care, risk protection, administrative reorganisation of small municipalities and telematic services, reduction of disparities between centralised urban areas and most fragile diffuse urban areas with high environmental risk.

The results of the work provide suggestions for risk protection of the most fragile diffuse residential systems in hilly and mountainous inland areas which can represent an alternative to the concentration and gathering of the population.

Finally, the paper proposes some operational strategic lines for a revision of the Urban Planning paradigms in the light of the Pandemic, adapting the most advanced disciplinary acquisitions in terms of protection from seismic risks to the wider needs of coexistence with pandemic dangers.

In conclusion, a rethinking of the widespread residential model is proposed, in synergy with the centralised residential model, where the need to avoid, in some temporal phases, concentrations of both people and densely built spaces, can be combined with the possibility of life in peripheral internal.

Keywords: Seismic and pandemic risks · Coexistence with risk · Fragile suburbs · Distancing-integration scenarios · Demographic concentration-diffusion

1 The Territorial Context

In Central Italy and, in particular, in the Marche, there is an unusual territorial characterisation of the peripheral areas (Bedini and Bronzini 2016): Green comet (a radiocentric systems of urban e rural fringes), Luminous serpentines (settlements over the hilltops of

© The Author(s), under exclusive license to Springer Nature Switzerland AG 2022
D. La Rosa and R. Privitera (Eds.): INPUT 2021, LNCE 242, pp. 143–151, 2022.
https://doi.org/10.1007/978-3-030-96985-1_16

Central Italy); Luminous paths (narrow lines of continuous settlements along the main valleys), Urban nebulae (chaotic constellations of scattered buildings).

In a time of economic recession, the transformation of these low density settlements might turn them into areas that attract investments and an experimental territory for the planning of innovation, social and equal-sustainable protection of rural-urban green areas, in an extensive landscape rich in social and ecologic interactions (Rauws and De Roo 2011). These urban structures, which are characterised by extreme and disorganised land use and a high risk of negative interaction, degradation and abandon, represent a new opportunity for a policy to relaunch local values (Stephenson 2010) and a way of life in environments with a low anthropisation level and high environmental landscape value: neither city, nor countryside, nor park. What is the future of these different typologies of peripheries? (Grant et al. 2013).

The research identifies numerous pathologies: the co-existence and competition of centrifugal and centripetal forces, settlement disorganisation without recognisability, identity, functionality, relationships between the parts; structural degradation including of the formal, functional and social quality of disintegrated urban settlements; a loss of the value of historical-cultural assets, dangerous interactions between vehicle and pedestrian traffic; an uncontrolled increase of costs per residential unit and the costs needed to resolve situations of unsustainable environmental incompatibility; a lack of functional organisation and rational management of the filiform settlement along the narrow urban sections, etc.

2   Objectives in the European Context

Some of the major objectives of the European Union (COM(2007) 621 2007) are:

- Mitigation of territorial competitions, which in the last decade has been established as competitiveness between institutions and territories, with the risk of degenerating into policies of "everyone against everyone", to the detriment of the weaker areas (low anthropisation areas), is replaced by a new form of competitiveness (based on the essence of the values in question and not on publicity form) between integrated city, countryside and fringe territory systems, as a binding driving force, which is being considered in a new way in terms of planning, design and management.
- Incentive of cohesion: the result of inconsistency (and poor quality of life), reached in the territories of modernity, is reversed by the expected results, in the sense of rethinking the economy as a real valuation of contexts, able to generate products of quality both in the field of agriculture (according to the natural vocation of the land) and in the entrepreneurial world, with the valuation of the irreplaceable resources of history, art, culture and landscape, ancient production systems that have been created in the world through to a very high development potential growth, ranging from cultural, environmental and gastronomy tourism to the green economy and new energy sources.

The *paper* intends to answer the question whether the new awareness and urban and territorial scenarios that emerged following the pandemic will lead to a transformation of the diffusion-demographic concentration relationship, both in terms of settlement structures and roles played on the one hand by the cities and on the other by the vast areas with low building density. These objectives are coherent with the provisions of the European Union and the requirements of the contents of the Recovery Plan.

Currently there are no known plans for regulating long settlements of isolated buildings, along the hill tops, the valley bottoms, and the expansion fringes of urban centers.

These new settlement systems involve many neighbouring and non-neighbouring municipalities and no attention is given to these settlement structures, seen in their supra-municipal development, that would on the other hand require decisive multi-scale planning. These low-density urban structures are characterised by extreme and disorganised land use and high risks of negative interaction (Mininni 2005; Nazio 2006).

The objectives of the Recovery Plan (Piano Nazionale di Ripresa e Resilienza 2021) can be summarised as: ecological transition, environmental regeneration, green economy, green city, expansion of digital activities. strengthening of widespread health care, streamlining of administrative procedures, and contrasting social inequalities and inequities between urban centers and small countries in inland areas, with high environmental risk.

3   Methodology and Research Results

In order to pursue the indicated objectives relating to risk assessment and mitigation strategies in urban and regional planning, it was decided to start from the reconsideration of the unstable balance of the relationship between suburbs, urban centers and urbanised countryside. We then proceeded with the identification of the seismic, hydrogeological and pandemic risks that threaten this balance to the detriment of widespread urbanised areas.

Subsequently, the research evaluates an overall strategy for the widespread residential model, supported by the resources made available by the European Union "Recovery Plan" for large strategic projects: ecological transition, digital transition, contrasting territorial disparity, administrative deburocratisation.

The research brought to light numerous pathologies and possible solutions at different scales of intervention with the aim of identifying, for each type of peripheral settlement, suggestions for their reinforcement and requalification.

Modalities for the reduction of the global risk are also identified, since the level of peripheralisation is strongly linked to the seismic and hydrogeological risk (Bedini and Marinelli 2021) which today, in the presence of a pandemic risk, poses new problems and potential.

The proposed solutions aim at overcoming the city-countryside dualism and controlling the development dynamics in the contamination between peri-urban and countryside settlements, in the landscapes in transit. The results of our research on these new settlement systems above all involve the serious shortcomings ascertained on the part of the public authorities in the management and planning of the territory.

In a phase of strong global change, even at a settlement level, with situations of abandon and degradation, a policy to relaunch local values and a way of life in environments with a low anthropisation level and high environmental landscape value, may represent a new opportunity for economic and social development.

In the context of the current pandemic, the small countries in inland areas and urbanised countryside will again be able to play a propulsive role, returning to a form of territorial balance with centralised areas, re-evaluating the role of the peripheries (linear, radiocentric, hilly serpentine, masses urbane of uncontrolled urban expansion). However, this situation can consolidate provided that these systems remain interconnected with the world (Tira 2020).

It is therefore time to rethink the management model of urban and rural space, reconstructing new hierarchical, functional and interchange relationships between center and periphery, between metropolitan areas and internal areas, with the generation of positive effects on the «interpersonal relationships of proximity» (Balducci 2020a).

However, only with access to European funding will it be possible to pursue the new proposed settlement model in concrete terms (Tarpino and Marson 2020).

Ultimately, from the ongoing debate emerges the need for urban planners and territorial planners to give their own contribution to redefine balanced and equitable social and territorial relationships between high residential density and residential scarcity, public space and private space, eliminating the major socio-spatial inequalities to urban, regional and national level (Pasqui 2019).

Finally, the global changes in the behavior of the population and in the settlement structures imposed by the pandemic leads to a mutation of the disciplinary paradigms of urban planning.

The pandemic will therefore cause a modification of the dynamic relationship between areas of demographic concentration and areas of demographic diffusion. This change in unstable equilibrium will affect the functions performed, both by densely populated cities and areas with scattered residences in inland areas. The strategies and contents of territorial and urban plans must therefore be reconsidered to define a different settlement model (Bedini and Bronzini 2018) where small countries located in inland areas can play a coherent role in symbiosis with the centralised settlement model, which is place in turn for safety.

Such a review of the planning tools starts from the assumption that the resources, made available to our country by the Recovery Fund, as part of the EU Next Generation plan for strategic projects to protect against global risks, constitute a unique and unrepeatable potential for concretizing the settlement structural model: internal areas, suburbs, urban centers.

Some changes to the planning contents can be summarised as follows:

- Definition of essential elements, at urban and territorial level, of Urban Minimum Pandemic Structures (SUMP).
- Realisation or reintroduction of neighborhood services and equipment in small inland urban areas.
- Computerisation of health services spread throughout the territory and forms of door-to-door assistance to businesses.
- Activation of high-tech equipped civil protection centers in the infrastructural nodes of inland areas, always accessible to first aid vehicles for mass health care.

- Planning of an urban-territorial structure that favors the consolidation of service hierarchies between urban centers and widespread urban areas (Bedini et al. 2019).
- Use of flexible and alternative land use destinations.
- Monitoring of buildings, at urban and territorial level, destined to accommodate functions required by post-seismic emergencies (Bedini and Bronzini 2018) or pandemic protection.
- Reactivation of infrastructures and equipment spread throughout the territory as a system of both safe mobility and emergency response in the event of a disaster.
- Creation of a territorial social and health protection system.
- Incentive to slow mobility.
- Variable delimitation of public spaces, squares, nightlife spaces, parks, to be inhibited or contingent upon in the event of a pandemic emergency.

Ultimately, the protection from the risks of the more fragile widespread residential systems of hilly and mountainous inland areas must be carried out in harmony with the safety of highly concentrated urban areas.

Some additions to the planning tools are now listed in line with the reforms required by the European Union, in the context of the Recovery Plan.

- Implementation of the concept of flexible, temporary and dynamic use of space, of variability of the times of the city, modifying the modalities of movement and access to urban facilities, social, cultural and sporting services;
- Enhancement of a symbiotic relationship between city, suburbs and countryside, for the reduction of territorial inequalities.
- Redesign of cities into more liveable neighborhoods, on a human scale.
- Creation of urban elements in countries of inland areas.
- Development of a "National and regional evacuation plan", in the event of natural pandemic disasters (and also of terrorist attacks, with conventional, chemical or biological weapons) which provides for a sudden reuse of the enormous underutilised or abandoned property assets available in internal areas.
- Monitoring of the unstable balance between the neighborhood function typical of widespread settlements and the propulsive function performed by urban centers. Monitoring of the dynamic relationships between central locations, widespread locations and linear interconnection peripheries.
- Creation of alert systems for residents and civil protection centers on the possible multiple consequences generated by earthquakes and hydro-geological quenches.
- Application of flexible zoning, with flexible uses of spaces, alternating between a state of everyday life and a state of emergency, at the municipal, urban and neighborhood level.
- Preventive preparation of historical-rural countries where to spend quarantine periods in emergency situations.
- Preparation of some small historical-rural centers such as Covid Free "islands".
- Setting up high-tech, protected and highly accessible centers located in a strategic nodal position serving wide areas.
- Reactivation, in internal areas, of small disused local hospitals as medical centers serving wide areas, construction of emergency centers, outpatient centers, strengthening of home medical care, telemedicine, school activities.

Ultimately, a revision of the times of the city is required (Zaoli 2020) in which to reorganize the city and a flexible use of the movement spaces (Monti 2020). Some initiatives (Milan, Bergamo, Prato, Bolzano, etc.) were also tested in implementation of Law 53/2000 or Legislative Decree 267/2000, with the rescheduling of the opening hours of businesses and public bodies (Tira 2020).

The "15-min cities" model (De Luca 2020) implies that each neighborhood is equipped with a set of neighborhood facilities and services, currently absent or abandoned (Balducci 2020a). The preventive delineation of spaces and functions and the planned distancing for risk protection was deepened in a city organisation model based on an integration between Epidemic Prevention Area (EPA) (Wei 2020) and "15 min city".

Therefore, scientific evidence from urban planning research shows that in the presence of a pandemic it is mandatory to proceed with the reorganisation of the city by neighborhoods, ensuring the presence of elements and values of proximity (Balducci 2020b). The activation or inhibition of attractive poles can be programmed, depending on the health and policy choices to combat the pandemic (Zaoli 2020). The Plans will therefore have to redesign the "cities of neighborhoods", where always ensure the autonomy of essential services, and the ability to reach, in fifteen minutes, even the bus or metro stations (Tira 2020). In other words, planning must continually recalibrate itself by adapting to the differences between planned guidelines and actual situation (Monti 2020). And programming must always remain interconnected with the ecological transition, the green economy, health and social assistance spread throughout the area.

In this logic of strong dynamism, the times and methods of drafting, approval, management of urban plans will also have to be completely modified (Tira 2020). And this also in accordance with the needs of acceleration, equity, finalisation, efficiency, flexibility, required by public administration reform and European Union for the allocation of the resources of the "Recovery Fund".

To counteract the inequalities between urban centers, suburbs and urbanised countryside with high environmental risk, policies to support more fragile areas must be supported (Ventura and Tiboni 2016), mitigating inequalities, not only between centralised areas and widespread settlements, but also between centers and suburbs and between north and south.

The small countries of inland areas will remain connected with a hierarchical structure of high technology health, energy, circular economy centers, protected from seismic, hydrogeological and pandemic risks. These villages will be able to make available, in case of emergency, many buildings and equipment, for the population forced to move from the cities.

Territorial planning will also be able to program the reuse of available resources, under or badly used, to favor cultural and food and wine tourism, modern agrozootechnics, the spread of creative artistic activities (Bedini and Bronzini 2019).

4   Final Remarks

The pandemic has caused the dichotomies of the pre-existing city to explode: the relationship between space and time, between contrast and coexistence with global risk, between immobile city and city in transformation in conditions of uncertainty, between densification and diffusion, between deterministic and strategic development guidelines, between rigidity of intended uses and flexibility of use. It follows the need to redesign the centralised urban model by creating green cities, where the need to avoid, in some temporal phases, concentrations of both people and densely built spaces, can be combined with the possibility of life in peripheral internal.

Ultimately, the mitigation of territorial fragilities and inequalities is only possible with a dynamic, unstable and symbiotic relationship between new functions of scattered settlements, capable of favoring qualitative neighborhood relations, and the indispensable role of cities, which in turn are secured in pandemic time.

In this perspective, the results of the research can be extended to other areas of the European Union with similar characteristics of the types of suburbs considered.

In conclusion, it could be said that long linear peripheral areas, hilly and valley floor, play a functional role in the territorial structural system, as a link between centralised areas and widespread rural areas. This role evolves on the basis of an unstable equilibrium, which can lead to spatial and a-spatial degradation in the event of the absence or abandonment of proximity services and production activities. Following hydrogeological instability and seismic events, the fragile unstable equilibrium is broken in the absence of strong and rapid funding, and management efficiency. Otherwise, the conditions of territorial imbalance and inequity cannot be removed. Finally, in pandemic situations, following the need for deconcentration, thinning out, isolation, the areas of settlement spread can be called upon to perform an important alternative life function and the suburbs (linear, radiocentric, hilly serpentine, urban masses of uncontrolled expansion) can regain their structural balance between urban centers and the urbanised countryside, if supported by large focused funding.

The pandemic, therefore, will certainly leave an indelible mark even in the paradigms of urban planning which will be called, in situations of uncertainty, to redesign cities, linear or aggregated suburbs, internal areas, by introducing flexible, temporary and uncertain uses. And the disciplinary *intellighenzia* will necessarily have to propose anti-pandemic strategies of urban and territorial policy, to reduce social and spatial inequities and differences in cities, regions and the Country.

To avoid such needs being accumulated in the limbo of good intentions without political and financial involvement, it must be clear that this opportunity can be seised immediately or never again in the context of the Recovery Plan.

References

Balducci, A.: I territori fragili di fronte al Covid [Fragile territories in front of the Covid]. Scienze del Territorio, special issue Living the territories in the time of Covid, pp. 169–176 (2020a). https://doi.org/10.13128/sdt-12352

Balducci, A.: Come cambiano le città dopo la pandemia [How cities change after the pandemic]. In: 28° Forum Scenari Immobiliari "Après le déluge", Santa Margherita Ligure, 11–12 Settembre (2020b)

Bedini, M.A., Bronzini, F.: Old and new paradigms in pre-earthquake prevention and post-earthquake regeneration of territories in crisis. Archivio di Studi Urbani e Regionali **124**, 70–95 (2019). https://doi.org/10.3280/ASUR2019-124004

Bedini, M.A., Bronzini, F.: The post-earthquake experience in Italy. Difficulties and the possibility of planning the resurgence of the territories affected by earthquakes. Land Use Policy **78**, 303–315 (2018). https://doi.org/10.1016/j.landusepol.2018.07.003

Bedini, M.A., Bronzini, F.: The new territories of urban planning. The issue of the fringe areas and settlements. Land Use Policy **57**, 130–138 (2016). https://doi.org/10.1016/j.landusepol.2016.05.020

Bedini, M.A., Bronzini, F., Marinelli, G.: Preservation and valorisation of small historical centres at risk. In: Gargiulo, C., Zoppi, C. (eds.) Planning, Nature and Ecosystem Services, pp. 744–756. FedOA Press, Napoli (2019). https://doi.org/10.6093/978-88-6887-054-6

Bedini, M.A., Marinelli, G.: Project suggestions for post-earthquake interventions in Italy. From building reconstruction to the population resettlement. TeMA **14**(1), 21–32 (2021). https://doi.org/10.6092/1970-9870/7568

COM(2007) 621: Communication of the European Communities, Agenda for a sustainable and competitive European tourism. Bruxelles (2007)

De Luca, G.: Il ruolo dello spazio pubblico come risorsa antipandemica [The role of public space as an anti-pandemic resource]. In: Nuovi paradigmi urbani e abitativi per le città post pandemia, Urbanpromo Green, Venezia, 18 Settembre (2020)

Grant, J.L., Nelson, A.C., Forsyth, A., Thompson-Fawcett's, M., Blais, P., Filion, P.: Planning Theory Practice **14**, 391–415 (2013)

Mininni, M.: Né città, né campagna. Un terzo territorio per una società paesaggista [Neither city, nor countryside. A third territory for a landscape society]. Urbanistica, vol. 128. Inu Edizioni, Rome (2005)

Monti, C.: Oltre la città razionalista: nuove prospettive e nuovi modelli urbani per il post pandemia [Beyond the rationalist city: new perspectives and new urban models for the post-pandemic]. In: Ingenio. Informazione tecnica e progettuale (2020). https://www.ingenio-web.it/29185-oltre-la-citta-razionalista-nuove-prospettive-e-nuovi-modelli-urbani-per-il-post-pandemia. ultimo accesso 17 Mar 2021

Nazio, P.: Lo spazio urbano-rurale [The urban-rural space]. In: Prestamburgo, S. (ed.) Sviluppo rurale in Europa [Rural development in Europe]. Urbanistica Informazioni, vol. 210. Inu Edizioni, Rome (2006)

Pasqui, G.: Il territorio al centro [The territory at the center]. Urbanistica Informazioni, vol. 287–288, pp. 10–11. Inu Edizioni, Roma (2019)

Piano Nazionale di Ripresa e Resilienza: Next Generation Italia. Roma (2021)

Rauws, W.S., De Roo, G.: Exploring transitions in the peri-urban area. Plan. Theory Pract. **12**, 269–284 (2011)

Stephenson, J.: People and Place. Plan. Theory Pract. **11**, 9–21 (2010)

Tarpino, A., Marson, A.: Dalla crisi pandemica il ritorno ai territory [From the pandemic crisis the return to the territories]. Scienze del Territorio, special issue Living the territories in the time of Covid, pp. 6–12 (2020). https://doi.org/10.13128/sdt-12369

Tira, M.: La pandemia come volano per il ripopolamento dei centri rurali? [How does the pandemic drive the repopulation of rural centers?] In: Samorì, C. (ed.) Ingenio. Informazione tecnica e progettuale (2020). https://www.ingenio-web.it/28124-la-pandemia-come-volano-per-il-ripopolamento-dei-centri-rurali. Accessed 17 Mar 2021

Ventura, P., Tiboni, M.: Politiche di sviluppo sostenibile per comunità urbane minori svantaggiate e conservazione del patrimonio naturale e culturale [Sustainable development policies for disadvantaged urban communities and conservation of natural and cultural herit-age]. In: Rotondo, F., Selicato, F., Marin, V., López Galdeano, J. (eds.) Cultural Territorial Systems. Landscape and Cultural Heritage as a Key to Sustainable and Local Development in Eastern Europe, pp. 29–49. Springer, Switzerland (2016). https://doi.org/10.1007/978-3-319-20753-7

Wei, D.: Urban Function-Spatial Response Strategy for the Epidemic. A Concise Manual on Urban Emergency Management, 18 Marzo (2020). https://www.ovpm.org/wp-content/uploads/2020/03/covid-19icomos-china.pdf. Accessed 20 Mar 2021

Zaoli, M.: L'urbanistica oltre l'emergenza del Covid 19: una città resiliente condivisa responsabile inclusive [Urban planning beyond the Covid 19 emergency: a resilient shared responsible inclusive city]. In: Blog Urbanistica Inu (2020). https://www.inu.it/assets/doc/urbanistica-e-covid-19-marco-zaoli.pdf. Accessed 17 Mar 2021

A Fragmentation-Based Analysis of Costa Viola (Southern Italy) Agricultural Terraces

Salvatore Praticò[1], Francesco Solano[2(✉)], Salvatore Di Fazio[1], and Giuseppe Modica[1]

[1] Dipartimento di Agraria, Università degli Studi Mediterranea di Reggio Calabria, Loc. Feo di Vito snc, 89122 Reggio Calabria, Italy
[2] Department of Agriculture and Forest Sciences (DAFNE), University of Tuscia, via San Camillo de Lellis, 01100 Viterbo, Italy
f.solano@unitus.it

Abstract. Traditional and historic agricultural landscapes cover a significant part of the Italian territory. For these landscapes, the study of change dynamics over a defined time interval is a crucial aspect to monitor their integrity and cultural value and, on the other hand, to set up appropriate strategies to assess their correct management. Terraced agricultural landscapes play an important role in this framework, given their substantial cultural and social value. Because of their artificial creation, this kind of landscape is subjected to continuous changes over time due to the interaction of natural and human factors. To detect the changes that occurred during a time interval of about 60 years, we adopted a fragmentation-based analysis of the terraced agricultural landscape of Costa Viola (Southern Italy). All analyses were conducted in a GIS environment through the free tool Landscape Fragmentation (LandFrag) and the period investigated goes from 1955 to 2014. We analysed the fragmentation of active terraces during three reference years: 1955, 1989 and 2014. A binary raster-based analysis has been conducted for each year according to five fragmentation classes: core, perforated, edge, transitional and patch areas. The dynamic highlighted by the analyses is characterised by the diminution of core areas, representing a stable landscape with no threat of fragmentation, that passed from 300.20 ha in 1955 to just 11.32 ha in 2014, demonstrating an important tendency towards fragmentation of the landscape leading to a gradual disappearance of traditionally cultivated terraces with an increased risk of landslides.

1 Introduction

By its definition, the landscape is constantly changing because it results from the interactions between the natural and human components [1, 2]. Rural landscapes are characterised on the one hand by a functional aspect, often related to land use/land cover (LU/LC) and, on the other hand, by a cultural aspect, related to the historical and cultural dimension of the communities that shape the landscapes to meet their needs [3]. Terraced agricultural landscapes play an important role in this framework, given their substantial cultural and social value. Traditional and historic agricultural landscapes cover a significant part of the Italian territory [4]. In European countries, during the last century, a

© The Author(s), under exclusive license to Springer Nature Switzerland AG 2022
D. La Rosa and R. Privitera (Eds.): INPUT 2021, LNCE 242, pp. 152–159, 2022.
https://doi.org/10.1007/978-3-030-96985-1_17

progressive abandonment of agricultural lands has been highlighted [5–8], especially in terraced systems [9–11], increasing landslide risk [12–15].

For these landscapes, the study of change dynamics over a defined time interval is crucial for monitoring their integrity and cultural value and setting up appropriate strategies to assess their correct management [1]. Several scholars have addressed the detection of terraced agricultural landscapes and their changes over time and the study of typological terraces features [9, 16–23]. LU/LC change analysis provides only a partial picture of the agricultural terrace dynamics. To better understand their evolution trends, it is crucial to detect the spatial configuration of terraced areas through the use of pattern change analysis. Landscape pattern change, i.e., landscape fragmentation, has been linked repeatedly with agricultural productivity and the threat of landscape and habitat loss [24–28]. This research aims to investigate active agricultural terraced system changes analysing their fragmentation dynamics in the study area of the Costa Viola, in the province of Reggio Calabria (Southern Italy) in a time period of about 60 years, from 1955 to 2014. To better understand the dynamics that occurred during this interval, 1989 was chosen as the intermediate year.

2 Materials and Methods

Study Area
Costa Viola, which literally means "purple coast," is a narrow coastal strip (1–2 km width) placed in the Tyrrhenian part of the province of the metropolitan municipality of Reggio Calabria (Southern Italy). It covers about 24 km^2, falling in the administrative territory of five municipalities: Villa San Giovanni, Scilla, Bagnara Calabra, Seminara and Palmi (from South to North - Fig. 1A). The area is characterised by an elevation between sea level and 600 m a.s.l., with very different slope classes (30°–45° is the most representative) and a prevailing North-North-West aspect orientation. Climate is typically the Mediterranean, with drought and hot summers and temperate winters. The terraced systems, dating from the 18th century onwards, are characterised by walls, built by specialised craftsmen, completely dry stone (Fig. 1B) and integrated into the anthropic urban environment (Fig. 1C). They are mainly used for vineyards, followed by olive orchards, citrus and to a lesser extent for horticultural crops.

LU/LC Digitalisation and Fragmentation Analysis
To detect the fragmentation dynamics of the active terraced systems of the Costa Viola, it was first necessary to precisely identify their position and shape in correspondence of each of the investigated years. This has been done, thanks to the characteristic signs of the dry stone walls seen from above, through photointerpretation and manual digitalisation of LU/LC in GIS environment. For 1955, historical B/W aerial photographs have been digitally processed and mosaicked into an orthophoto. For 1989, a digital B/W aerial orthophoto available as a web map service (WMS), by the Geoportal of the Italian Ministry of Environment, has been used as a reference layer for the digitalisation process. For 2014, a WorldView-2 satellite-based image has been used. Moreover, to detect the terraced system, a fourth hierarchical level has been implemented to CORINE Land Cover legend for the agricultural areas. More details about photogrammetric and digitalisation processes can be found in our previous studies [9, 20].

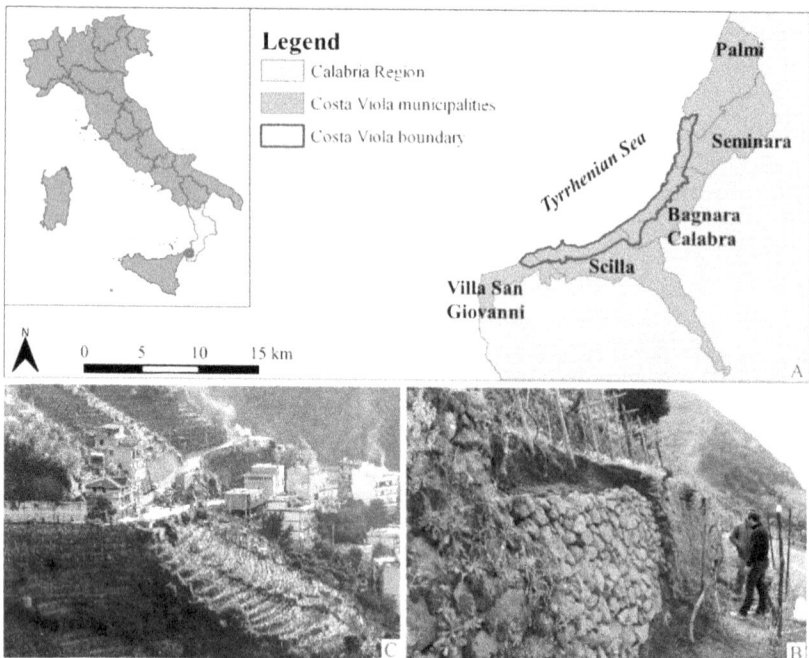

Fig. 1. The Costa Viola study area.

All pattern analyses were conducted in a GIS environment through the free tool Landscape Fragmentation (LandFrag) [29]. A binary raster-based analysis has been conducted for each year, rasterising the obtained LU/LC maps. For all active agricultural terraced LU/LC classes the value 1 was assigned, 0 for all other LU/LC classes (Fig. 2). A 3 × 3 moving window has been adopted to detect the 8-neighbourhood pattern of the central pixel, according to five fragmentation classes: i) core, which is the most stable situation in which the central pixel is part of an active terraced area and it is entirely surrounded by pixels with the same LU/LC (active agricultural terraces); ii) perforated, more than 60% of pixels of the moving window belong to an active terraced area and defines the boundary between core and its small perforations; iii) edge, more than 60% of pixels of the moving window belong to an active terraced area and defines the boundary between core and its large perforations or the boundary between active terraces core and other LU/LC areas; iv) transitional, between 60% and 40% of the moving window's pixels are classified as active terraces; v) patch, is the least stable and most worrying situation, representing areas that are active terraces but they are too small to contain core areas.

3   Results and Discussions

According to the results of the digitalisation, the amount of the area occupied by active agricultural terraces passed from 813.25 ha in 1955 to 302.59 ha in 1989 and finally decreased to 118.79 ha in 2014 (Fig. 2).

As highlighted in previous studies [9, 20, 30] this abandonment trend, mainly due to a more general abandonment of agricultural activities jointly to a cadastral parcels dissolution process, has mainly affected the most sloping and difficult-to-reach areas.

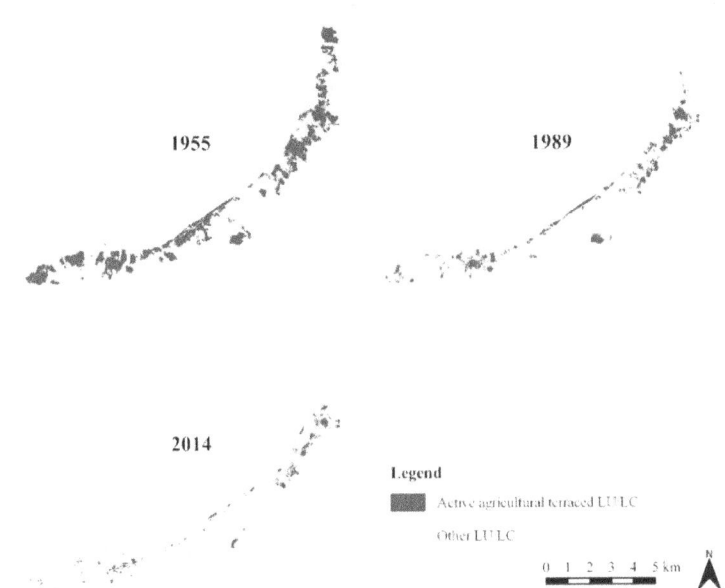

Fig. 2. Active agricultural terraced LU/LC maps for each investigated year (1955, 1989 and 2014).

Due to the different extensions of the active terraced areas in the analysed years, the fragmentation analysis results are reported in percentage terms for each class. Results of fragmentation analysis are synthesised in Fig. 3, while Fig. 4 shows fragmentation classes maps.

Core areas, the most stable pattern, registered a continuous decrease passing from 36.94% in 1955 to 21.02% in 1989 and 9.59% in 2014. The decrease of the core areas is combined with the increase of the patch areas that passed from 2.87% in 1955 to 10.82% in 1989 and 22.87% in 2014. These two trends align with the highlighted agricultural terraced areas abandonment and the more general trend of abandonment of agricultural areas. In addition, the dynamic of transitional areas, which increased from 11.81% in 1955 to 21.07% in 1989 and 26.43% in 2014, could be read as a sign of the progressive abandonment of agricultural practices on terraced fields. Analysing the maps in Fig. 3, the same decreasing trend can be highlighted for the core and perforated areas. Instead, the transitional and patch areas share the same increasing trend. Edge areas registered an increase between 1955 and 1989 (25.30% to 27.05%) and decreased between 1989 and 2014 (27.05% to 23.31%).

These results show how the active agricultural terraces, due to fragmentation processes mainly driven by anthropogenic factors, have become less stable over time and therefore more in need of attention. Moreover, the obtained spatial localisation of these areas makes it possible to identify the most critical ones, in order to calibrate the management actions and, where necessary, facilitate access to public funds.

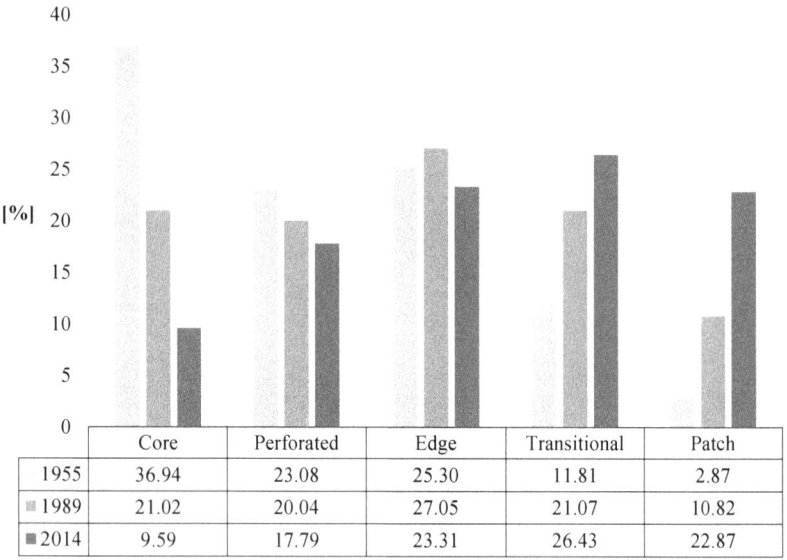

Fig. 3. Percentage occupied by analysed fragmentation classes

	Core	Perforated	Edge	Transitional	Patch
1955	36.94	23.08	25.30	11.81	2.87
1989	21.02	20.04	27.05	21.07	10.82
2014	9.59	17.79	23.31	26.43	22.87

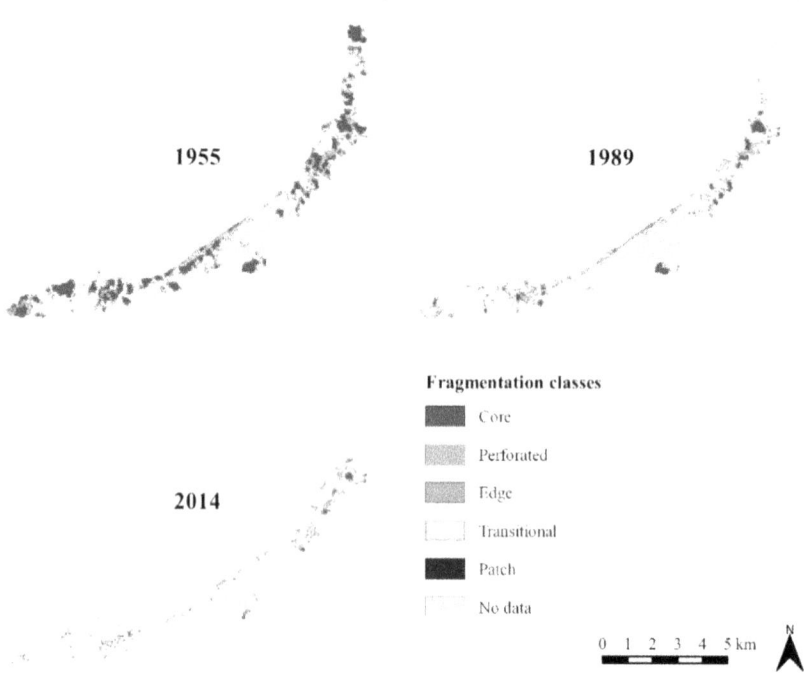

Fig. 4. Fragmentation class maps for each investigated year (1955, 1989, and 2014).

4 Conclusion

In this study, we analysed the dynamics of the agricultural terraces of the Costa Viola area, focusing on their pattern changes through a multitemporal raster-based fragmentation analysis. The authors are aware of the possible limitations and constraints of the proposed method that may arise from the use of images from different periods and sources, especially with regard to photointerpretation process, because of their different spatial and spectral resolution. The study highlighted the abandonment of terraced agricultural landscapes and showed the fragility of the remaining terraced areas that are more vulnerable to the fragmentation dynamics that lead to abandonment, thus triggering a dangerous loop. An intelligent and well-calibrated aid program should be put in place to stop this phenomenon, ensuring that funding can be accessed to preserve agricultural activities in these particular traditional landscapes.

References

1. Antrop, M.: Why landscapes of the past are important for the future. Landsc. Urban Plan. **70**, 21–34 (2005). https://doi.org/10.1016/j.landurbplan.2003.10.002
2. Council of Europe: European Landscape Convention (2000)
3. Di Fazio, S., Modica, G.: Historic rural landscapes: sustainable planning strategies and action criteria. The Italian experience in the global and European context. Sustain **10**, 1–27 (2018). https://doi.org/10.3390/su10113834
4. Agnoletti, M.: Italian historical rural landscapes: dynamics, data analysis and research findings. In: Agnoletti, M. (ed.) Italian Historical Rural Landscapes. ENVHIS, vol. 1, pp. 3–87. Springer, Dordrecht (2013). https://doi.org/10.1007/978-94-007-5354-9_1
5. Modica, G., Vizzari, M., Pollino, M., Fichera, C.R., Zoccali, P., Di Fazio, S.: Spatio-temporal analysis of the urban-rural gradient structure: an application in a Mediterranean mountainous landscape (Serra San Bruno, Italy). Earth Syst. Dyn. **3**, 263–279 (2012). https://doi.org/10.5194/esd-3-263-2012
6. Lasanta, T., Arnáez, J., Pascual, N., Ruiz-Flaño, P., Errea, M.P., Lana-Renault, N.: Space–time process and drivers of land abandonment in Europe. CATENA **149**, 810–823 (2017). https://doi.org/10.1016/j.catena.2016.02.024
7. Ramankutty, N., Foley, J.A.: Estimating historical changes in global land cover: croplands from 1700 to 1992. Glob. Biogeochem. Cycles **13**, 997–1027 (1999). https://doi.org/10.1029/1999GB900046
8. MacDonald, D., et al.: Agricultural abandonment in mountain areas of Europe: Environmental consequences and policy response. J. Environ. Manage. **59**, 47–69 (2000). https://doi.org/10.1006/jema.1999.0335
9. Modica, G., Praticò, S., Di Fazio, S.: Abandonment of traditional terraced landscape: a change detection approach (a case study in Costa Viola, Calabria, Italy). L. Degrad. Dev. **28**(8), 2608–2622 (2017). https://doi.org/10.1002/ldr.2824
10. Lasanta, T., Errea, M.P., Nadal-Romero, E.: Traditional agrarian landscape in the mediterranean mountains. a regional and local factor analysis in the central Spanish pyrenees. L. Degrad. Dev. **1640**, 1626–1640 (2017). https://doi.org/10.1002/ldr.2695
11. Lieskovský, J., et al.: The abandonment of traditional agricultural landscape in Slovakia – analysis of extent and driving forces. J. Rural Stud. **37**, 75–84 (2015). https://doi.org/10.1016/j.jrurstud.2014.12.007

12. García-Ruiz, J.M., Lana-Renault, N.: Hydrological and erosive consequences of farmland abandonment in Europe, with special reference to the Mediterranean region – a review. Agric. Ecosyst. Environ. **140**, 317–338 (2011). https://doi.org/10.1016/j.agee.2011.01.003
13. Llorens, P., Latron, J., Gallart, F.: Analysis of the role of agricultural abandoned terraces on the hydrology and sediment dynamics in a small mountainous basin (High Llobregat, Eastern Pyrenees). Pirineos **139**, 27–46 (1992). https://doi.org/10.3989/pirineos.1992.v139.180
14. Moreno-de-las-Heras, M., et al.: Hydro-geomorphological consequences of the abandonment of agricultural terraces in the Mediterranean region: key controlling factors and landscape stability patterns. Geomorphology **333**, 73–91 (2019). https://doi.org/10.1016/j.geomorph.2019.02.014
15. Romero Díaz, A., Marín Sanleandro, P., Sánchez Soriano, A., Belmonte Serrato, F., Faulkner, H.: The causes of piping in a set of abandoned agricultural terraces in southeast Spain. CATENA **69**, 282–293 (2007). https://doi.org/10.1016/j.catena.2006.07.008
16. Capolupo, A., Kooistra, L., Boccia, L.: A novel approach for detecting agricultural terraced landscapes from historical and contemporaneous photogrammetric aerial photos. Int. J. Appl. Earth Obs. Geoinf. **73**, 800–810 (2018). https://doi.org/10.1016/j.jag.2018.08.008
17. Demoulin, A., Bovy, B., Rixhon, G., Cornet, Y.: An automated method to extract fluvial terraces from digital elevation models: the Vesdre valley, a case study in eastern Belgium. Geomorphology **91**, 51–64 (2007). https://doi.org/10.1016/j.geomorph.2007.01.020
18. Agnoletti, M., Conti, L., Frezza, L., Santoro, A.: Territorial analysis of the agricultural terraced landscapes of Tuscany (Italy): preliminary results. Sustain **7**, 4564–4581 (2015). https://doi.org/10.3390/su7044564
19. Di Fazio, S., Modica, G.: The valorisation and characterisation of the agrarian terraced landscape. A case study in the Costa Viola area (Italy). In: International Conference of Agriculture Engineering CIGR-AgEng (2012)
20. Modica, G., Praticò, S., Pollino, M., Di Fazio, S.: Geomatics in analysing the evolution of agricultural terraced landscapes. In: Murgante, B., et al. (eds.) ICCSA 2014. LNCS, vol. 8582, pp. 479–494. Springer, Cham (2014). https://doi.org/10.1007/978-3-319-09147-1_35
21. Wei, W., et al.: Global synthesis of the classifications, distributions, benefits and issues of terracing. Earth Sci. Rev. **159**, 388–403 (2016). https://doi.org/10.1016/j.earscirev.2016.06.010
22. Tarolli, P., Sofia, G., Calligaro, S., Prosdocimi, M., Preti, F., Dalla Fontana, G.: Vineyards in terraced landscapes: new opportunities from lidar data. L. Degrad. Dev. **26**, 92–102 (2015). https://doi.org/10.1002/ldr.2311
23. Lanucara, S., Praticò, S., Modica, G.: Harmonization and interoperable sharing of multi-temporal geospatial data of rural landscapes. In: Calabrò, F., Della Spina, L., Bevilacqua, C. (eds.) New Metropolitan Perspectives. ISHT 2018. SIST, vol. 100. Springer, Cham (2019). https://doi.org/10.1007/978-3-319-92099-3_7
24. Penghui, J., Dengshuai, C., Manchun, L.: Farmland landscape fragmentation evolution and its driving mechanism from rural to urban: a case study of Changzhou city. J. Rural Stud. **82**, 1–18 (2021). https://doi.org/10.1016/j.jrurstud.2021.01.004
25. Rahman, S., Rahman, M.: Impact of land fragmentation and resource ownership on productivity and efficiency: the case of rice producers in Bangladesh. Land Use Policy **26**, 95–103 (2009). https://doi.org/10.1016/j.landusepol.2008.01.003
26. Modica, G., Praticò, S., Laudari, L., Ledda, A., Di Fazio, S., De Montis, A.: Implementation of multispecies ecological networks at the regional scale: analysis and multi-temporal assessment. J. Environ. Manag. **289**, 112494 (2021). https://doi.org/10.1016/j.jenvman.2021.112494

27. Heider, K., Rodriguez Lopez, J.M., Balbo, A.L., Scheffran, J.: The state of agricultural landscapes in the Mediterranean: smallholder agriculture and land abandonment in terraced landscapes of the Ricote Valley, southeast Spain. Reg. Environ. Change **21**(1), 1–12 (2020). https://doi.org/10.1007/s10113-020-01739-x

28. Solano, F., Praticò, S., Piovesan, G., Chiarucci, A., Argentieri, A., Modica, G.: Characterising historical transformation trajectories of the forest landscape in Rome's Metropolitan area (Italy) for effective planning of sustainability goals. L. Degrad. Dev. ldr. **32**(16), 4072 (2021). https://doi.org/10.1002/ldr.4072

29. Vogt, P., Riitters, K.H., Estreguil, C., Kozak, J., Wade, T.G., Wickham, J.D.: Mapping spatial patterns with morphological image processing. Landsc. Ecol. **22**, 171–177 (2007). https://doi.org/10.1007/s10980-006-9013-2

30. Arnáez, J., Lana-Renault, N., Lasanta, T., Ruiz-Flaño, P., Castroviejo, J.: Effects of farming terraces on hydrological and geomorphological processes. A review. CATENA **128**, 122–134 (2015). https://doi.org/10.1016/j.catena.2015.01.021

Sustainable Planning of the Rural Landscapes in the Northwest of the Iberian Peninsula: Best Practices and Management Conservation

Ignacio J. Diaz-Maroto$^{(\boxtimes)}$

Department of Agroforestry Engineering, Higher Polytechnic School of Engineering, University of Santiago de Compostela, Campus Terra s/n, 27002 Lugo, Spain
ignacio.diazmaroto@usc.es

Abstract. The conservation of rural landscapes and traditional forest practices are key aspects to achieve sustainable rural development. The dynamics of the landscape is the result of anthropological action over several centuries ago. We have studied the development and the environmental, socioeconomic, and historical changes have taken place. Our aim is to propose a set of actions of landscape planning for the conservation and recovery of the traditional uses, i.e., for to encourage sustainable management of the cultural landscapes. Little attention has been given to the resilience of these areas. Therefore, we have considered it essential to analyse the transformation of these landscapes in the mountains of eastern Galicia, considered as a (complex) socio-ecological system. The study focused on the evolution of native forests, intensively exploited since prehistoric times. These ecosystems were converted into agricultural land, exploited for the naval, metallurgical and railway industries, joined the Church's properties, suffered forest fires, and were replaced by fast-growing species, mainly coniferous and now *Eucalyptus nitens* Shining Gum. All these activities have led to a decrease in the area occupied by them. Now, native forests cover small and usually sloping sites, remaining where soil properties often avoid other land-use. There is an altered landscape by a slow change and the biodiversity conservation, hunting and environmental tourism have great value. From a positive view, the area covered by these forests has increased and there is a greater perception on their conservation given the recognition as habitats of European interest –Natura 2000 Network.

Keywords: Land-use · Sustainability · Cultural landscape · Biodiversity · Native forests

1 Introduction

Global environmental changes present exceptional challenges to nature and humanity. It arises the need for an interdisciplinary understanding of relationships between society and natural formations. Socioenvironmental history overlaps humanity and the ecology through different processes of change in time and space to clarify the complex story of development. Current challenges associate to changes in socioenvironmental relationships, the acceleration of deforestation, trafficking of fauna species, pollution and global

© The Author(s), under exclusive license to Springer Nature Switzerland AG 2022
D. La Rosa and R. Privitera (Eds.): INPUT 2021, LNCE 242, pp. 160–167, 2022.
https://doi.org/10.1007/978-3-030-96985-1_18

warming are some examples (Van Eetvelde and Antrop 2004; Scheidel et al. 2020). In our scale of work, these challenges could compromise to achieve a sustainable rural development (Manuel and Gil 2001).

In this way, the variety of the cultural landscapes is frequently bigger than in the natural or seminatural landscapes submit on the heterogeneity created by nature and anthropoid activities. Traditional land-use in the Eastern Galician Mountains –northwester of the Iberian Peninsula– has shaped a system controlled by seasonal cycles and spatial models of human activities (Diaz-Maroto and Vila-Lameiro 2008). Now, socioeconomic globalization has driven multiple changes and uncertainties in the agricultural, forestry and landscape services of rural areas that affect their future, however trivial attention has been paid to their resilience. The land abandonment has reduced the open spaces and has generated an expansion of the forests by impacts for both ecosystems and biodiversity. This means decrease of wide-open habitats, obstacle to the agropastoral activities, and ever-increasing wildfires. Furthermore, an increase in forest area is likely to the loss of open-habitat and ecotone species (Buide et al. 1998). The interaction between people and the environment plays a vital role in the configuration of the landscapes, mostly in those exposes to human impact where socio-ecological relationships regulate biodiversity (Farina 2000; Burel and Baudry 2001).

Despite the historical reduction of the forest area took place until recent times, due to a set of factors such as: farming and food production, grazing, unsuitable forest management, confiscation of forests owned by the Church, forest fires, and afforestation by fast-growing species; now, the area of these forests has increased. Also, there is a major awareness of the importance of their conservation given the recognition as habitats of European interest –Natura 2000 Network–. It is a network of natural areas aims to make the protection of species and habitats compatible with human activity, promoting the good conservation status (Galicia Government 2018).

During the last times, socioeconomic globalization has caused deep changes in rural areas (Ahern 1994). Globalization can be defined as the process of international integration which has economic, social, and political dimensions (Dreher 2006). Many countries have adapted to this process and have the benefit of the welfare effects of globalization by implementing necessary economic and institutional transformation. However, some countries still suffer from poor adaptation to global markets, being necessary to promote a suitable rural development. Sustainability depends largely on the preservation of traditional uses, in our case, extensive grazing and sustainable forest management (Sanderson and Harris 2000). Extensive grazing a suitable number of cattle and avoiding their concentration in small areas, prevents erosion and vegetation decline, increases diversity, and maintains open-habitat spaces. Selective logging is a viable economic activity and integrated environmentally aiding to reduce forest fires due to landscape fragmentation (Debussche et al. 1999).

The environmental, cultural, and economic mixing of agrosilvopastoral systems seems essential (Farina 2000) to ensure the conservation of these landscapes. Our aim is to propose a set of actions for the conservation and recovery of the traditional uses. For that, we analysed the resilience of the rural landscapes in the mountains of eastern Galicia, because little is known about the extent to which the concept of resilience can be applied to rural development. The study focused on the progression-regression of native forests, intensively exploited since ancestral times.

2   Resilience of Landscapes in the Eastern Mountains of Galicia

2.1   Study Area: Western Mountains of the Cantabrian Range

The Mountains of Eastern Galicia, Ancares-Courel range is a transitional area from typical Atlantic to Mediterranean flora (Diaz-Maroto and Vila-Lameiro 2007). It ranges at the western of the Cantabrian Mountains (Fig. 1). It is an area of sloped land and the elevations fluctuating between 250 and nearby of 2,000 m. The climate is typified by rainfall varying between 700 mm in the lower areas and nearly 2,500 mm in the tops. The annual temperature regime is very severe with long winters in the highlands. Annual average temperatures vary from 4.6 °C to 14.0 °C in the most protected locations (Ramil et al. 2013).

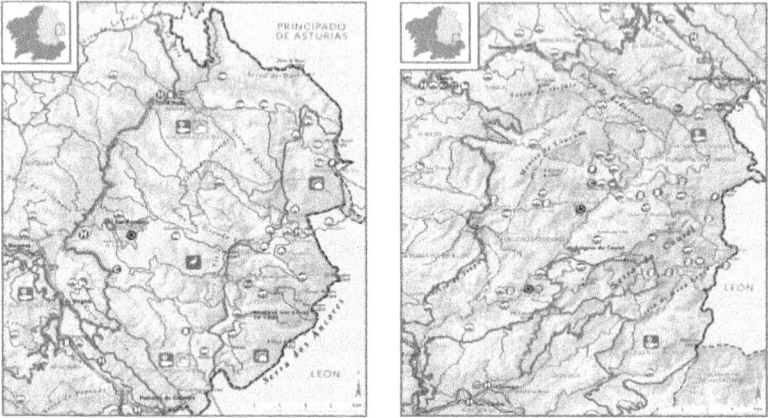

Fig. 1 Western mountains of the Cantabrian range (northwest Spain) (Source: www.turgalicia.es)

The potential vegetation should at present occupy this region would be the native broadleaf forests, characterized by different species of genus *Quercus* (Buide et al. 1998). Agreeing to different research, these forests were established in the study area between five and seven thousand years ago. Chronological-historical factors, site requirements and needs of each species gave rise to different type of forests and floristic formations (Peterken and Game 1984).

2.2   Agrosilvopastoral System: Function and Dynamics

The exploitation system in the study area is an agrosilvopastoral organization adapted to the heterogeneousness of the mountainous environment, based on cereal crops, especially rye, and extensive grazing with orchard crops and fruit trees (Diaz-Maroto and Vila-Lameiro 2008). This socioeconomic model has shaped the cultural landscapes (Sobrino et al. 2001). Farmland and mowing meadows are located at the bottom of valleys and the villages in slope middle with sunny orientation. Orchards and fruit trees placed between the houses surround the villages and give way to chestnut stands *"soutos"* (Manuel and Gil 2001). The scrubs dominate the steep sunny slopes, where even periodic burning is carried out to regenerate the mountain pastures. On the shady slopes the human impact is very low, being refuge from the best examples of primitive forests, mixed forests with presence of distinct species of Atlantic oaks (Manuel and Gil 2001; Diaz-Maroto and Vila-Lameiro 2007).

Fig. 2 Agrosilvopastoral system founded on the maintenance of traditional uses: extensive agropastoral activity and suitable forest management (Diaz-Maroto, 2020-07-20)

Many of these natural ecosystems (Fig. 2), were replaced by fast growing species, mainly conifers and now *Eucalyptus nitens* Shining Gum. The Autonomous Adminis-tration has recently approved a moratorium on new eucalyptus plantations.

2.3 Sources: Chorological and Historical Information

The chorological data allows us to analyse the land-use changes because to anthropo-logical action. The value of this information has been highlighted by several authors (Chocarro et al. 1990; Van Eetvelde and Antrop 2004). The chorological info enables the vegetal ecosystems origin and the natural changes that happen in their allocation to be well-defined (Marris 2007).

Data about the land alterations because to human activities was obtained from several sources. As our research aims on the latter, historical data was completed with more recent information included in the Annual Agrarian Statistics (from 2000 to present), National Forest Inventories (NFIs), Forest Plan of Galicia of 1992, and the first revision of the Forest Plan of 2018. All databases were necessary for to achieve our objective because each offers a distinct type of information. For example, the Forest Plan of Galicia and its revision showed the increase in the natural broadleaved forests during the last few decades is likely the result of the natural dynamic but may also be affected by other factors such as the depopulation and abandonment of rural land.

3 Cultural Landscapes of the Galician Eastern Mountains: Sustainable Planning and Management Best Practices

Cultural landscapes could be defined as areas with properties that linked works of nature and humanity. The interdisciplinary character of sustainable planning and management of them demands nature and culture conservation strategies are integrated into one holistic system, where the preservation of cultural landscapes embrace both the key issues of nature conservation and preservation of biodiversity, sustainable use of ecosystem services and the conservation of built and intangible cultural heritage (Lagerqvist et al. 2016). Also, an evolutionary approach should be applied seeing the constant changes of landscape and their use. Conservation must be able to integrate preserved natural-cultural values into development strategies by defining their use and role in society. Ecosystem services play a key part in conservation of landscapes since they constitute the basis of strategic use of natural resources for sustainable landscape planning. An ecosystem-centred holistic management structure which will enable landscape planning to protect rural destiny from overexploitation and design touristic level according to the social-ecology carrying capacity (Lagerqvist et al. 2016).

3.1 Expansion of Native Broadleaved Forests: Beginning of Agriculture and Grazing, First Signs of Deforestation

Landscape changes involve a long period of dominance by *Quercus* forests (Fig. 3). Exploitation timber is carried out for centuries because of the plenty of high-quality wood. Despite this, Forestry Administration policy have proposed fast-growing species use –pine and eucalyptus. This fact played an important role in shaping the landscape, although changes in the production system has been a key factor (Crecente et al. 2002). Modifications in the rural landscape were not large until agriculture and grazing became global (Santos et al. 2000). Agriculture began in Galicia over 5500 years ago, during the megalithic culture (Torras et al. 1980). The first farmers were nomads who removed vegetation by fire before to cultivate grain until production declined and left that land. The livestock accelerated soil degradation (Pons et al. 2003). During the Middle Age, several factors driven to reduction in the forest area until the second half of the 19th century for later to giving rise to constant growth. Afforestation use was significant after the Spanish Civil War and during the end 20th century (Bauer 1980).

3.2 Consequences for the Conservation of Native Forests

Now, the native broadleaf forests area has significantly increased (Fig. 3). According to the IV National Forest Inventory, these forests occupy a 31% of the forest area, 441,289 ha, standing out the area covered by *Quercus robur* L., 246,446 ha –17.4%– of the forest cover (Galicia Government 2018). Most oak forests lack of management because to their limited economic interest, but have a high environmental relevance being habitats of Community interest as part of the Natura 2000 Network. As we said, Natura 2000 is a European network that aims to protect natural and semi-natural species and habitats. It is also the world's largest network of protected areas (Skliar et al. 2019).

Fig. 3. Galician Oriental Mountains: native broadleaf forests (Diaz-Maroto, 2020-07-14)

3.3 Dynamics and Forecast of the Rural Landscapes in the Study Area

Landscape dynamics has been marked by progression-regression periods of the native forests. Across of the time, the regression cycles have been wider and more intensive, except in latest decades, where has it taken place an expansion of natural ecosystems (Santos et al. 2000). To assess the conservation current conditions as well as the future management forecast, historical processes must be known. Depending on the natural-cultural effects the landscape changes are seen over time (Jaeger 2000; Gökyer 2013).

There are two main factors of landscape change. These are the natural processes and human actions. Both natural conditions and human activities are changed over time. Complex transformations can be developed in the landscape structure linked to natural environment and human needs (Farina 2000). Landscapes change naturally as they are the expression of the interaction between the environment and human's activities.

4 Conclusions

The presence of broadleaved forests in the NW of the Iberian Peninsula has been associated with changes in land-use as result of human influence. The processes involved in the reduction of natural broadleaved forests have been long and complex. They have implicated wars, invasions, felling for shipbuilding and steel industries, change to agricultural use, construction of railways, charcoal making, and the afforestation with pines and eucalyptus. These facts gave rise to a decrease in the area occupied by these formations until the middle of the 19th century. In the second half, there was a trend change, and there is an increase in the area occupied by these species.

Rural demands are continually subjected to change, including modifications in agriculture, extensive cattle farming, firewood production, and exploitation of timber for the naval, iron and railway industries. Afforestation played a key role in shaping the forest landscape, particularly in relation to fast growing species, involving critical changes in the productive system. The changes that have taken place in recent decades explain the distribution of current cultural landscape as a common system of rural management (agriculture-forestry-grazing) adapted to the mountain environmental.

This productive model is completed with a predominance of small owners and a tendency for agriculture to be abandoned. The property is often inherited by town or city inhabitants with no interest in agrarian-forest practices. This drives the emergence of unproductive land with the consequent risk of forest fires occurrence. The implementation of silvicultural practices with the aim of improving and restoring native forests would enable the recovery of the cultural landscape as part of the natural-environmental heritage. This would minimize the effects of forest fires and of the increase of unproductive land and maximize diversification of obtained forest products, improve biodiversity, and stimulate the interest of the new owners.

References

Ahern, J.: Greenways as ecological networks in rural areas. In: Cook, E.A., van Lier, H.N. (eds.) Landscape Planning and Ecological Networks, pp. 159–177. Elsevier, Amsterdam (1994)

Bauer, E.: Los Montes de España en la Historia. Ministerio de Agricultura, Madrid (1980)

Buide, M.L., Sánchez, J.M., Guitián, J.: Ecological characteristics of the flora of the Northwest Iberian Peninsula. Plant Ecol. **135**, 1–8 (1998)

Burel, F., Baudry, J.: Ecologie du Paysage: Concepts, Méthodes et Applications. Éditions Tec & Doc-Lavoisier, París (2001)

Crecente, R., Alvarez, C.J., Fra, U.: Economic, social, and environmental impact of land consolidation in Galicia. Land. Use Pol. **19**, 135–147 (2002)

Chocarro, C., Fanlo, R., Fillat, F., Marin, P.: Historical evolution of natural resource use in the central Pyrenees of Spain. Mt. Res. Dev. **10**, 257–265 (1990)

Debussche, M., Lepart, J., Dervieux, A.: Mediterranean landscape changes: evidence from old postcards. Global Ecol. Biogeogr. **8**, 3–15 (1999)

Dreher, A.: Does globalization affect growth? Evidence from a new index of globalization. Appl. Econ. **38**, 1091–1110 (2006)

Diaz-Maroto, I.J., Vila-Lameiro, P.: Deciduous and semi-deciduous oak forests (*Quercus robur, Q. petraea* and *Q. pyrenaica*) floristic composition in the Northwest Iberian Peninsula. Biologia **62**, 163–172 (2007)

Diaz-Maroto, I.J., Vila-Lameiro, P.: Historical evolution and land-use changes in natural broadleaved forests in the north-west Iberian Peninsula. Scand. J. Res. **23**, 371–379 (2008)

Farina, A.: Landscape Ecology in Action. Springer, Dordrecht (2000)

Galicia Government: First review of the Forestry Plan of Galicia: Diagnosis Report of the Forests and the Galician Forest Sector. Ministry of Rural Environment, Santiago de Compostela (2018)

Gökyer, E.: Understanding landscape structure using landscape metrics. In: Özyavuz, M. (ed.) Advances in Landscape Architecture, pp. 663–676. IntechOpen, London (2013)

Jaeger, J.A.: Landscape division, splitting index, and effective mesh size: new measures of landscape fragmentation. Landsc. Ecol. **15**, 115–130 (2000)

Lagerqvist, B., et al.: Sustainable management of cultural landscapes: use of ecosystem services in rural tourism. In: Beskidzka, S. (ed.) Ecological Footprint in Central Europe, pp. 91–107. The University College of Tourism and Ecology Press, Kraków (2016)

Manuel, C.M., Gil, L.: La Transformación Histórica del Paisaje Forestal en Galicia. Ministerio de Medio Ambiente, Madrid (2001)

Marris, E.: Linnaeus at 300: The species and the specious. Nature **7133**, 250–253 (2007)

Peterken, G.F., Game, M.: Historical factors affecting the number and distribution of vascular plant-species in the woodlands of central Lincolnshire. J. Ecol. **72**, 155–182 (1984)

Pons, P., Lambert, B., Rigolot, E., Prodon, R.: The effects of grassland management using fire on habitat occupancy and conservation of birds in a mosaic landscape. Biodivers. Conserv. **12**, 1843–1860 (2003)

Ramil, P., Rodríguez, M.A., López, H., Ferreiro da Costa, J., Muñoz, C.: Loss of European dry heaths in NW Spain: a case study. Diversity **5**, 557–580 (2013)

Sanderson, J., Harris, L.D.: Landscape Ecology A Top-Down Approach. Lewis Publishers, by CRC Press LLC, Boca Raton, Florida (2000)

Santos, L., Romani, J.R.V., Jalut, G.: History of vegetation during the Holocene in the Courel and Queixa Sierras, Galicia, northwest Iberian Peninsula. J. Quat. Sci. **15**, 621–632 (2000)

Scheidel, A., et al.: Environmental conflicts and defenders: a global overview. Glob. Environ. Change **63**, 102104 (2020)

Skliar, V., Kovalenko, I., Skliar, I., Sherstiuk, M.: Vitality structure and its dynamics in the process of natural reforestation of *Quercus robur* L. AgroLife Sci. J. **8**, 233–241 (2019)

Sobrino, C.M., Ramil-Rego, P., Guitián, M.R.: Vegetation in the mountains of northwest Iberia during the last glacial-interglacial transition. Veg. Hist. Archaeobotany **10**, 7–21 (2001)

Torras, M., Díaz-Fierros, F., Vázquez, J.: Sobre el comienzo de la agricultura en Galicia. Gallaecia **6**, 51–59 (1980)

Van Eetvelde, V., Antrop, M.: Analyzing structural and functional changes of traditional landscapes—two examples from southern France. Landsc. Urban Plan. **67**, 75–95 (2004)

The Monumental Heritage in Sardinia by Historical Eras: A Research to Evaluate the Place-Based Connections

Chiara Garau$^{(\boxtimes)}$ ⓘ, Giulia Desogus ⓘ, Alfonso Annunziata ⓘ, Federica Banchiero ⓘ, and Pasquale Mistretta ⓘ

Department of Civil and Environmental Engineering and Architecture (DICAAR), University of Cagliari, 09129 Cagliari, Italy
cgarau@unica.it

Abstract. The purpose of this paper is to investigate in a preliminary way the monumental heritage in Sardinia (Italy), formed in the different historical eras (the early-medieval, the medieval and the Catalan-Castilian periods), in order to understand how today the cultural landscape formed by these eras, is a potential attraction for residents and tourists. A model on cultural landscape applied to Sardinia Region is applied for obtaining this goal. This model is based on comparison between the integration indicator that analyses the accessibility, and the recreation index that interprets the use of the cultural good by users. The representation of the different historical periods sees the monuments as a place that history has strongly marked with traces and signs, sometimes disappeared and sometimes still clear and visible. The final goal of this paper will therefore be to propose a method of analysis of the cultural potential of locations across the Sardinian Region, through the accessibility and use analysis, which can be useful to the administration to create a representative network where the aggregated monumental realities can be part of different cultural paths also as tourism attractors. This paper shows an extension of a research on the paths of history for multicultural tourism started in 2019.

Keywords: Monumental heritage · Integration indicator · Recreation index · Depthmap X software · InVEST software · Sardinia region

1 Introduction

Nowadays, the urbanization process is increasing significantly (ONU 2019) by changing the spatial distribution in cities and, for that reason, outlining a clear distinction between todays and oldest areas of attraction. These areas are delineated by different relationships between natural, socio-cultural, economic and complex urban systems (Pintus et al.

[i] This paper is the result of the joint work of the authors. In particular, the "abstract", "Methodology and case study of Sardinia (Italy)", and "Results" were written jointly by all authors. Giulia Desogus wrote the "Introduction" and Alfonso Annunziata wrote the "Conclusions". Chiara Garau and Pasquale Mistretta supervised the paper.

© The Author(s), under exclusive license to Springer Nature Switzerland AG 2022
D. La Rosa and R. Privitera (Eds.): INPUT 2021, LNCE 242, pp. 168–174, 2022.
https://doi.org/10.1007/978-3-030-96985-1_19

2019; McKercher, Du Cros 2002; Lowenthal 2005; Vecco 2010; Cicinelli et al. 2017; Tengberg et al. 2012). Indeed, the distribution and subsequent use of cultural heritage are factors that determine, within the cities and its surrounding, areas of attraction both for residents and tourists. According to (Pintus et al. 2019) place-based policies have to identify procedures, guidelines, methods by which to manage these areas of attraction, "through a support to ensure 1) the access of the places to visit and of transport means, 2) correct travel times at the beginning and end of the routes, 3) stops for refreshments" (Pintus et al. 2019, p. 743). This would help both to guarantee an active governance of cultural heritage and an economic and occupational induced, and to fully satisfy the visitor demand of the good.

Over the years, the Italian government has issued several regulatory decrees on cultural heritage (Law 1939a; Law 1939b; Council of Europe 2000). However, the connection between regional planning and cultural heritage has underlined since the early 2000s with the Legislative Decree no. 42/2004, that introduces the Code of Cultural Heritage and the Environment (known as Urban Code). In it the content of urban planning at a national level was redefined, setting specific environmental quality objectives for each territory according to recognized value levels (Garau and Pavan 2010).

Nevertheless, the enhancement of the cultural landscape is still an open debate, especially if it is related to a regional territory and not to a single city. In this sense, the link between cultural goods, as single elements, and their connection with the current urban fabric becomes fundamental for the evaluation and planning of smart routes based on the use and the current accessibility of the cultural good. Indeed, cultural goods as single elements can become a strong point in the regional system by activating programmed connection and facilitated access methods. In this sense, they can become points of attraction in the regional tourist system which would allow greater use with economic and social advantages for the resident population (Garau 2014).

Starting from these assumptions, this study primarily determines an approach to analyse cultural landscape by considering a particularly sensitive context, such as Sardinia. Sardinia appears to be particularly emblematic as a case study because i) it has a plurality of types of assets present and ii) these goods are historical memories of different territorial contexts (Pintus et al. 2019).

The working hypothesis is to investigate the connection between accessibility and use of cultural heritage located on the island. In this regard, the intent is to pre-sent a hypothesis for analysing the territory, aimed at enhancing not only the good as such, but together with the user perception studied in different historical eras, in order to understand how these different eras have privileged some areas that in turn, today, have become areas of attraction. The analysis is made on the cultural heritage related to the early-medieval period (VI-XI centuries), the medieval period (XII-XIV), and the Catalan-Castilian period (XV-XVIII centuries). This choice is due both to the availability of data within the Landscape Regional Plan of the Sardinia Region and to the history of Sardinia which sees its greatest construction of cultural heritage in these times. The paper is therefore organized as follows: Sect. 2 shows the methodology and its relation to the case study of the Sardinia Region, Sect. 3 shows the results. Finally, Sect. 4 discusses the results and the future development of the research.

2   Methodology and Case Study of Sardinia (Italy)

This paper focuses on the cultural potential in different historical eras of Sardinia Island, one of the largest islands in the area of the Mediterranean basin with an area of 23,813 square kilometers. The island management of cultural heritage makes the region an emblematic case study (Council of Europe 2000; Melis et al. 2018). In fact, Sardinian Region, as a special administrative area, has the task of exercising the general planning functions of the regional cultural system, implementing the principle of vertical and horizontal subsidiarity. In compliance with the general guidelines established by the Sardinian Region, the Provinces, the Municipalities and the Metropolitan Cities (Cagliari e Sassari) will exclusively exercise the planning and management functions on the territory of the cultural goods (Fig. 1).

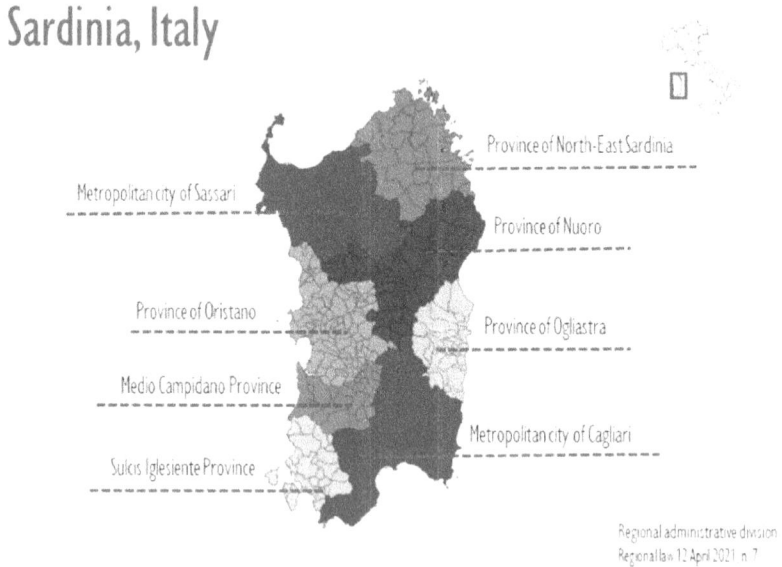

Fig. 1. Regional administrative division

For local development, the research process with practical application starting from the regional data of the Landscape Regional Plan (LRP) is crucial for the assessment of the relationship between spatial planning and cultural complexity (Pintus et al. 2019). This would help the local administration to define an easy and effective governance formula to start a system project which includes the whole island.

With these premises, the study proposes a method of analysis of the cultural potential of locations across the Sardinian Region, conceptualized as the combination of density of tangible components of the cultural heritage, and of conditions of access. This is because in literature regional paths for the entire Sardinian region don't exist for a specific era. The authors try to understand if 1) the interest of tourists in the cultural goods of specific eras exists; 2) today's accessibility allows to create them.

This study is the preliminary stage of a research on the cultural landscape of the Sardinia Region. The analysis considers sites and elements of cultural interest ascribable to three distinct periods: the early-medieval period (VI-XI centuries), the medieval period, characterized by the influence of the Republic of Pisa (XII-XIV), and the Catalan-Castilian period (XV-XVIII centuries). Data, related to cultural points of interest are retrieved from the Landscape Regional Plan of the Sardinia Region. The unit of analysis considered is the cell of a regular square grid. Density of cultural components is estimated via a Kernel Density estimation, considering an area of influence determined by a 5,000 m radius.

Accessibility, is conceptualized as the potential of a location as destination, and it is measured by the Integration indicator, calculated via the Depthmap X software. Values of density and of integration are normalized and aggregated into the Cultural Potential Index. A Recreation Index, is calculated via the InVEST Software, and is measured as the total photo-user-days for each grid cell. One photo-user-day at a location is one unique individual who posted at least one picture on a specific day. For each cell, the model sums the number of photo-user-days for all days from 2005 to 2017, and returns the average annual number of photo-user-days. Lastly, the correlation among the Cultural Potential Index and the Recreation Index is measured via the R^2 coefficient of determination.

3 Results

Results reveal a specific spatial pattern in the distribution of sites of cultural relevance: in particular, sites ascribable to the XII-XIV centuries are concentrated in the northern part of the Sardinia Region, and Sites of cultural relevance related to the XV-XVIII centuries are observed in the eastern area of the Sardinia Region (Fig. 2, sx).

A further spatial pattern is related to the distribution of conditions of access, and it underlines the centrality of the metropolitan area of Cagliari in the road system of Sardinia. The combination of density sites of cultural relevance and of conditions of access determines the emergence of a polarized structure of sites of relevant cultural-touristic potential, in the western area of Sardinia, and of a dispersed, isotropic structure in the eastern part: in particular, the urban areas of Cagliari, Oristano, Bosa, Alghero and Sassari, emerge as relevant nodes (Fig. 2, dx).

The analysis of the recreational potential of the Sardinia Region (Fig. 3), reveals a dispersed structure of localities of recreational relevance.

In particular, the distribution of values of the average number of photo-user-day, underlines the urban areas of Cagliari, Bosa, Alghero, Oristano, Sassari, and the system of touristic localities along the coastal area. More precisely, significant values of the recreational potential indicator are related to touristic sites including Chia, Villasimius, Carloforte. As a result, the corelation analysis reveals a modest co-relation among values of the cultural potential indicator and the recreational potential indicator, measured by a R^2 coefficient equal to 0.29.

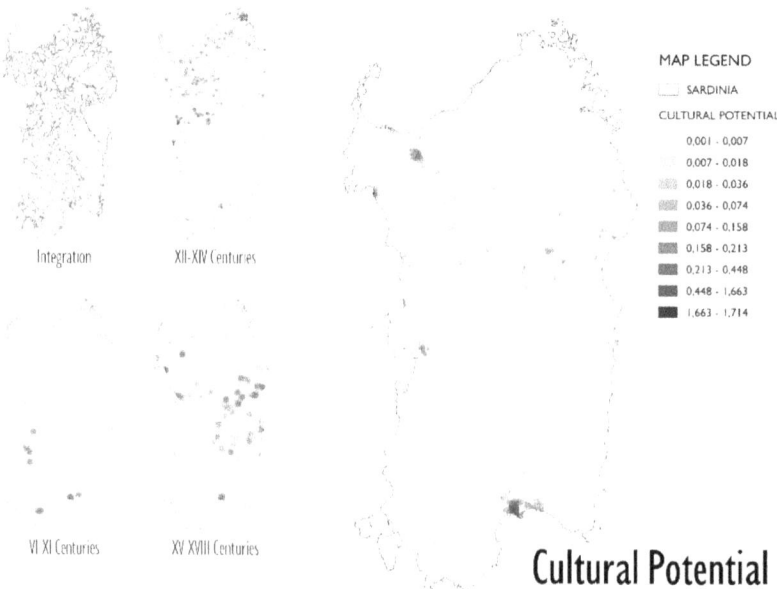

Fig. 2. Spatial scheme on distribution of sites of cultural relevance in different Eras (sx) and polarized structure of sites of relevant cultural-touristic potential (dx)

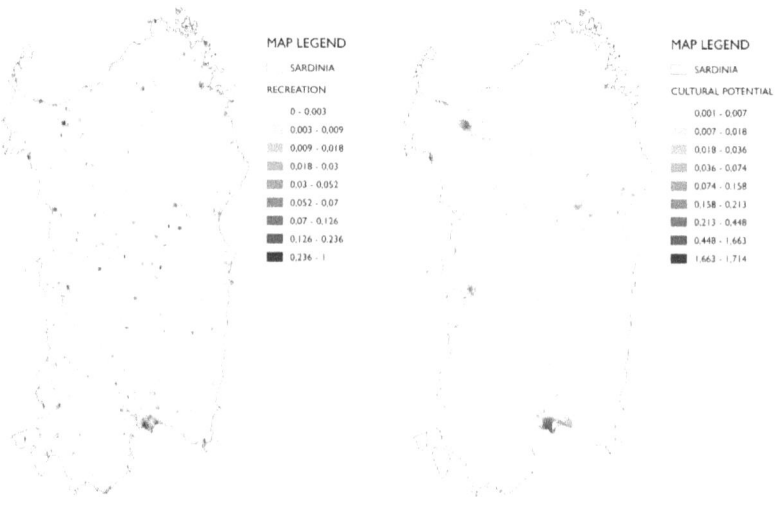

Fig. 3. Significant values of the recreational potential indicator

4   Discussion and Conclusions

The research first has outlined a descriptive model on the cultural landscape applied to Sardinia Region based on three different historical eras (the early-medieval period, the medieval period and the Catalan-Castilian period). Subsequently, through this model has studied a system of relationships that expands throughout the region. This is analysed through the comparison between the Integration indicator that allows to study the potential of a location as destination (accessibility), and the Recreation Index that interprets the use of the good by users through the averages annual number of photo-user-days.

In this way, proposed research outlines an explanatory framework of how much cultural heritage in Sardinia are used in relation to their accessibility. This has led to the development of a model capable of understanding the dynamics existing between accessibility and cultural heritage. In addition, the definition of thematic paths allows a better fruition of cultural goods that currently doesn't exist.

An observation concerns the modest impact of cultural landscapes in the Sardinia tourism industry and the need to promote alternative forms of tourism, in particular eco-tourism and cultural tourism, reinforcing the centrality of sites of cultural relevance located in inner regions of the Sardinia Island.

As a result, the development and utilization of the proposed model can support regional planning in two ways. First, by promoting the understanding of the evolution of settlement structures, via the description of the distribution of tangible components of cultural heritage related to specific eras; second, by facilitating the identification of relevant cultural sites marginal in terms of recreational and touristic utilization. These findings, combined with the recognition of relevant natural areas, and of available infrastructural resources, constitute the information base for the development of planning actions aimed at improving conditions of access, reinforcing services and reinforcing synergies among environmental, cultural resources, valorisation of specialities and promotion of productions relevant in cultural and environmental terms.

Indeed, the analyses made in this paper would allow local politics to confirm or modify the accessibility and the typology of services to be spread in the territory in favour of the use of cultural heritage.

For these reasons, future research will be oriented in two complementary directions: the first one envisages extending the methodology to other Historical Eras in order to have a complete framework for to encourage administrative actions (at all levels) for development of whole Island. The second one involves the study of actions aimed at improving the overall efficiency of the management of cultural heritage in the region, both with studies that favour the improvement and implementation of regional database data, and with actions that increase the accessibility of cultural heritage in a perspective of smart governance and smart tourism.

Acknowledgments. This study was also supported by the MIUR (Ministry of Education, Universities and Research [Italy]) through the project WEAKI TRANSIT: WEAK-demand areas Innovative TRANsport Shared services for Italian Towns (Project protocol: 20174ARRHT_004; CUP Code: F74I19001290001), financed with the PRIN 2017 (Research Projects of National Relevance) programme. We authorize the MIUR to reproduce and distribute reprints for Governmental purposes, notwithstanding any copyright notations thereon. Any opinions, findings and conclusions or recommendations expressed in this material are those of the authors, and do not necessarily reflect the views of the MIUR.

References

ONU World Urbanization Prospects. World Urbanization Prospects: The 2018 Revision (2018). https://population.un.org/wup/Publications/Files/WUP2018-KeyFacts.pdf

Pintus, S., Garau, C., Mistretta, P.: The paths of history for multicultural tourism: a smart real world in the metropolitan city of Cagliari (Italy). In: 24th International Conference on Urban Planning and Regional Development in the Information Society, Karlsruhe Institute of Technology, Germany, 2–4 April (2019)

McKercher, B., Du Cros, H.: Cultural tourism: The partnership between tourism and cultural heritage management, Routledge (2002)

LowenthaL, D.: Natural and cultural heritage. Int. J. Heritage Stud. **11**(1), 81–92 (2005)

Vecco, M.: A definition of cultural heritage: from the tangible to the intangible. J. Cultural Heritage **11**(3), 321–324 (2010)

Cicinelli, E., Salerno, G., Caneva, G.: An assessment methodology to combine the preservation of biodiversity and cultural heritage: the San Vincenzo al Volturno historical site (Molise, Italy). Biodivers. Conserv. **27**(5), 1073–1093 (2017). https://doi.org/10.1007/s10531-017-1480-z

Tengberg, A., Fredholm, S., Eliasson, I., Knez, I., Saltzman, K., Wetterberg, O.: Cultural ecosystem services provided by landscapes: assessment of heritage values and identity. Ecosyst. Serv. **2**, 14–26 (2012)

Law 1089/1939. Tutela delle cose d'interesse Artistico o Storico, Gazzetta Ufficiale (1939)

Law 1497/1939. Protezione delle bellezze naturali, Gazzetta Ufficiale (1939)

Council of Europe. European Landscape Convention (2000). https://rm.coe.int/1680080621

Garau, C., Pavan, V.: Regional cultural heritage: new vision for preservation in Sardinia (Italy). J. Landsc. Stud. **3**, 127–138 (2010)

Garau, C.: Smart paths for advanced management of cultural heritage. Reg. Stud. Reg. Sci. **1**(1), 286–293 (2014)

Melis, G., Zamperlin, P., Deguy, P., Garau, C.: Il censimento dei beni cut rali sul territorio regionale sardo, Innovazioni e potenzialità. In: XXII Conferenza Nazionale ed EXPO 2018 ASITA, pp. 677–684. Federazione ASITA (2018)

Smart and Slow Tourism. Evaluation and Challenges in Sardinia (Italy)

Ginevra Balletto[1]([✉]), Giuseppe Borruso[2], Mara Ladu[1], and Alessandra Milesi[1]

[1] DICAAR - Department of Civil and Environmental Engineering and Architecture, University of Cagliari, Via Marengo 2, Cagliari, Italy
{balletto,mara.ladu}@unica.it, alessandramilesi@gmail.com
[2] DEAMS - Department of Economics, Business, Mathematics and Statistics Sciences "Bruno de Finetti", University of Trieste, 34127 Trieste, Italy
giuseppe.borruso@deams.units.it

Abstract. Slow tourism is a sustainable way of traveling: it is not based on the consumption of resources and at the same time is linked to the discovery of local places and traditions. Furthermore, it favors a harmonious and responsible development of the territory in contrast with over tourism. Within this scenario and in the wake of a tradition rooted in Europe, the paths have become one of the most successful declinations of slow tourism. In this sense, the paths are increasingly becoming an integral part of the portfolio of tourism products in Sardinia, together with other types of tourism related to boating, culture, archeology, villages, cycling, food and wine, nature. The aim of this work is to investigate the role of slow tourism in Sardinia, with particular reference to the qualitative/quantitative evaluation of the model of the path of Santa Barbara Sulcis-Iglesiente and its exportability in other local contexts to enhance landscapes through the ancient tangible and intangible of mine habitats. Furthermore, the authors analyze the main potentially replicable elements of the path of Santa Barbara also in order to define best practices for the planning of slow tourism associated with smart tourism.

Keywords: Slow tourism · Smart tourism · Sustainable tourism · Sardinia

1 Introduction

Italy has a very ancient tourist tradition thanks to the variety of its landscapes [1] and the rich historical, artistic and natural heritage of global significance [2, 3], confirming itself as the nation that holds the largest number of sites included by UNESCO in the list of world heritage sites (55) [4, 5].

This work is part of the activities of the scientific committee of the 'Foundation Mining Path of Santa Barbara' (prof G. Balletto, Prof R. Paci, Prof I. Meloni, University of Cagliari). The paper is the result of the shared reflections, research and work of the authors involved. However, Mara Ladu and Alessandra Milesi realized paragraph 1 and paragraph 2; Ginevra Balletto and Giuseppe Borruso paragraph 3 and paragraph 4.

© The Author(s), under exclusive license to Springer Nature Switzerland AG 2022
D. La Rosa and R. Privitera (Eds.): INPUT 2021, LNCE 242, pp. 175–182, 2022.
https://doi.org/10.1007/978-3-030-96985-1_20

Although in the last twenty years the tourism sector has experienced exponential growth on a global scale, the health emergency marked a sharp break in that trend that seemed destined to increase in the coming years. In fact, the Covid-19 pandemic has hit the whole world, putting all economic sectors in crisis. In this context, the tourism sector is among the most vulnerable, being affected both in terms of supply and demand [6]. The imposed restrictions lead to recognize the sustainable tourism, in its declination of slow tourism, as the most responsive to new needs [7]. The 2017 has been defined as the International Year of Sustainable Tourism by the UN, which recognized its specific role in pursuing inclusive economic growth in balance with the environment. In Italy too, after the national year of the paths (2016) 2019 has been declared the year of slow tourism [8], understood as that form of experiential tourism, innovative and sustainable capable of enhancing the territory in all its diversity as it makes movement, even more than halfway, an integral part of the journey [9]. The paths are becoming the first symbol of sow tourism in Italy [10, 11] and a strategic way to promote little known contexts, especially in the internal areas of the country [12, 13].

The present study investigates the role of slow tourism, in particular of the walks, in defining the policies for the tourism development of the Sardinia Region and takes the territory of Sulcis-Iglesiente as a research area, where the recent establishment of the Santa Barbara Path - Mining Route in a sensitive area marked by intense mining activity, offers the opportunity to set up an initial assessment of the effects of these initiatives in terms of enhancement of the landscape and of development of the local community. The conception of the mining landscape and the surrounding heritage as an element to be valued originated with the establishment of the Geo-mining Historical Environmental Park of Sardinia (Geo-mining Park) in 2001 to safeguard and enhance tangible and intangible values in specific areas of the Region affected by a past extractive activities.

After an initial assessment of the geo-mining contexts and their enhancement, the authors analyze the principles and potentially replicable elements and demonstrate how this first experience is proving to be a best practice for starting a structured planning of walkability at the regional level, also involving other territorial contexts that they choose to promote forms of slow tourism starting from the enhancement of the local environmental, historical and cultural resource.

The present study is organized into the following parts:

– Material (paragraph 2), which describes the system of Mining Paths of Santa Barbara in Sardinia, Italy;
– Methodology (paragraph 3), which aims to develop a comparative index, applied to the case study;
– Study results and conclusions (paragraph 4).

2   Toward a System of Mining Paths of Santa Barbara in Sardinia, Italy

The economy of Sardinia, for a long time, has been characterized by an intense mining activity that has bequeathed a landscape of abandoned sites and industrial archeology which, in recent decades, has been at the center of a series of policies and projects aimed at its progressive enhancement. A first action was the establishment of the Geo-mining Historic and Environmental Park of Sardinia (Geo-mining Park), as a tool to protect the geological context, the cultural heritage of industrial archeology, traditions and knowledge. It extends to the regional level, including the main territorial areas affected by extractive activities. In fact, it consists of 8 areas covering 81 Municipalities of Sardinia, for a total area of approximately 3800 km^2 [14].

In the wake of this awareness, in Sulcis-Iglesiente, a historic region of south-western Sardinia, which was the most important district for national and international mining due to its large production of lead and zinc, has been interested by bottom-up initiatives for the protection and enhancement of the territory. First, the birth of the Pozzo Sella Onlus Association in 2001, then, the creation of the Santa Barbara Path and the establishmentof the Foundation called "Cammino Minerario di Santa Barbara" in 2016, with the task of implementing, promoting and managing the path [7, 15–19].

All these initiatives represent a response to the economic crisis in the mining sector and to the subsequent closure of the mines in the 1990s, which left a rich heritage of industrial archeology and infrastructure, as well as a unique landscape.

After the positive experience of the Santa Barbara Mining Path (01_MP) of the Sulcis-Iglesiente and in the light of the growing interest developed in it, the Pozzo Sella Onlus has launched an intense activity aimed at verifying the existence of favorable conditions for the construction of new walks in other island contexts characterized by a similar historical, cultural, environmental and religious heritage inherited from the mining activity developed in Sardinia.

These analyzes have led to the development of the proposal for the establishment of three other itineraries dedicated to the patron saint of miners in order to enhance the former paths linked to mining present in other areas of Sardinia (Fig. 1):

- 02_MP in the north-west, in the Nurra area
- 03_MP in the central south in the territories of Ogliastra, Barbagia Seulo and Sarcidano
- 04_MP in the south-east in the Sarrabus Gerrei area.
    The common features of the proposed itinerary refer to:
- Total length (km) along a ring route and altimeters (m);
- Stages/days of walking (km per stage);
- Paths, dirt roads, asphalted roads and/or paving in built-up areas (%);
- Stages and stopping points coinciding with the abandoned mines to be regenerated and with other sites equipped with existing accommodation facilities and/or to be implemented.

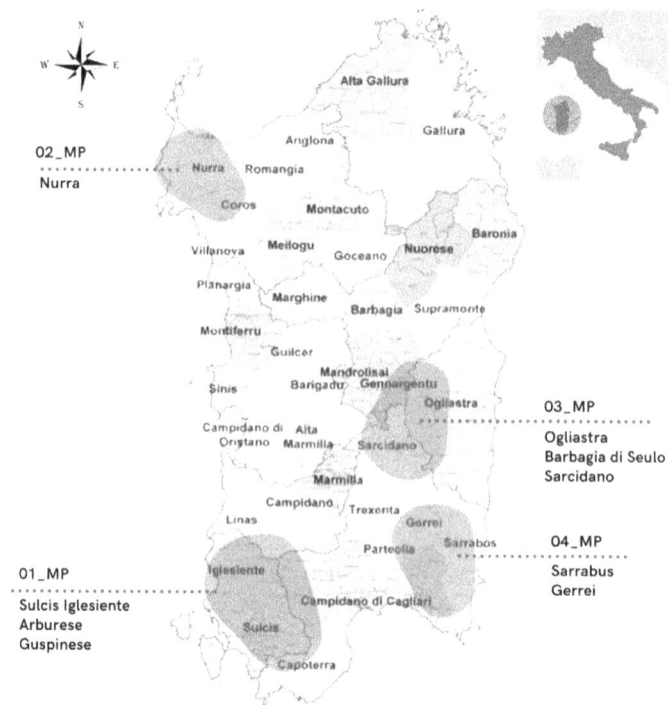

Fig. 1. The four mining paths of Santa Barbara, Sardinia, Italy (01–04 MP)

3   Methodology: Proposals for a Comparative Index

The following methodology aims to compare the recently proposed Santa Barbara paths, here named 02, 03 and 04 MP, with the first realized Santa Barbara path, here named 01_MP. As a matter of fact, the 01_MP represents a 'best practice' [7, 21] in terms of export and replication of the mining path model for several reasons. The presence of a significant number of municipalities involved in the project constitutes the prerequisite for the construction of more complex networks. At the same time, a greater number of inhabitants and of the territorial area of incidence represent decisive aspects for guaranteeing a better offer of cultural services and tourist accommodation for the users of the path. Finally, the greater presence of pre-existing itineraries constitutes the basis for the development of place-based projects, capable of enhancing the signs, shapes and characters of the landscape.

Within this framework, the authors developed a composite index referring to the inclusion of community and environment (ICE Index).

In particular, the methodology allows to evaluate how the recent proposals of the new three paths (02–04 MP) are in line with the Guidelines for the development of the Territorial Tourist Product, Santa Barbara Mining Trail [22], with particular reference to the following factors: prevention, through the enhancement of oriented projects to anticipate social and environmental problems; promotion, through community-environment policies based on well-being and quality of life; participation, through an active and responsible role in terms of the community and the environment; partnerships, with network forms for shared planning and joint management of initiatives. In this sense, the main factors of inclusion (community-environment), such as belonging to the Geo-mining Park (an area with strong characterizations in terms of enhancement of the mine landscape; number of municipal administrations involved, local reference population, incidence of natural-original mine paths,…) constitute the first indicators on which the ICE index is based. The ICE index aims to compare proposals and to support decisions for the advancement of their design. The conceptual model of the method is shown below. It will be subject to future application on a larger number of indicators that can be extrapolated from a database built by the authors in previous research [7] and by the Foundation of the Santa Barbara path itself [20]. In particular, the ICE index can be represented as follow:

$$\text{ICE Index} = \frac{\sum_{i=1}^{n} i * Wi}{\sum_{i=1}^{n} Wi}$$

where i = incidence factor of community-environment inclusion factors (divided into prevention, promotion, participation, partnerships). The definition of specific weights, which also derive from 'Guidelines for the development of the Territorial Tourist Product, Santa Barbara Mining Route' [see note 2], is particularly important in the computation of the different indexes. ICE index must then be divided by maximum Ice index in order to obtain ICE index relating to the group for which the comparison is to be made.

The objective of this contribution is to define criteria for the comparative analysis of the Santa Barbara Path in the Sulcis-Iglesiente area with the three new paths being defined in other historical regions of Sardinia. According to the Santa Barbara Path Foundation, the comparative analysis is based on those aspects that are gradually emerging as positive factors for the definition of the new paths. Since the four paths analyzed fall within the Geo-mining Park, it was considered appropriate to evaluate, for each one: the incidence in terms of territorial area concerned and number of municipalities crossed within the same Park, the population and number of municipalities involved, as well as the percentage incidence of original and/or rural routes.

The authors then proceeded to extrapolate the input data, and then elaborate the ICE index (absolute and relative value) referring to the 'best practice' (01_MP). It should be noted that wi = 1 / n, where n = number of incidence factors (i), in accordance with the Statute of the Santa Barbara Path Foundation and with the first outcomes of the Strategic Plan of the homonymous route.

In the summary scheme (Fig. 2) the inputs and outputs are reported. The first results that emerged are discussed in Sect. 4.

Path name	N. Municipalities sub area of the Geo-mining park	N. Municipalities crossed by the path	Municipalities areas crossed (sq km)	N inhabitants of the municipalities crossed	% of original or natural paths	ICE index absolute value	ICE index relative value
01_MP Sulcis Iglesiente Arburese Guspinese	39	26	2 416	153 076	75%	38887	1
02_MP Nurra	5	6	1 020	97 387	60%	24604	0.63
03_MP Ogliastra Barbagia di Seulo Sarcidano	4	5	420	6 623	35%	1763	0.04
04_MP Sarrabus Gerrei	11	9	874	29 184	75%	7519	0.19
Total	59	46	4730	386 270			

Fig. 2. Framework - input data and output data

4 Results

The three proposals of paths 02, 03 and 04 MP constitute a first territorial application of replication with adaptation of the 'best practice' 01 MP. In this sense, the evaluation of the ICE index (Fig. 2) allows us to highlight how the three proposals still differ considerably from the reference case, in particular 03_MP. In fact, the few municipalities and the small size of the communities involved determine an ICE index about half lower than that of the reference case (01_MP).

In light of this assessment, in the design phase of the three paths (02–04 MP) it will be necessary to build solid networks of active participation of communities and institutions also outside the system, in order to achieve similar conditions with respect to the reference case. For example, new parts of the path could be included in the 02_MP project that allow strengthening the connection with other settlement, cultural and environmental centralities present in the territorial context.

In addition, even the incidence of the originality of the route - very low in 03 MP - can be recovered through a slow multimodal mobility system. In other words, the exportability of the best practice, at the moment, does not occur in absolute terms for the following reasons: the proposed routes insist in marginal portions compared to the sub-areas of the Geo-minig Park, and therefore do not benefit from the related direct and the limited number of municipalities involved does not facilitate network actions. In fact, at the basis of the success of the paths, in addition to material actions, the ability to network between institutions and communities plays a strategic role, able to favor the transition from anticommons (mining activities) to commons (museum, cultural tourism activities), through semicommons (temporary cultural and tourist use) [23]. In the process of defining 03_MP it could be useful to involve more municipalities, also considering the possibility to connect the two ring paths 03_MP and 04_MP, by virtue of their geographical proximity.

The next phases of the research will be aimed at defining further elements of incidence (i) to support the evaluation of the ICE (Inclusion of Community and Environment) index, organized by macro-categories (prevention, promotion, participation, partnerships) for which the weights will be determined in accordance with the three-year update of the Strategic Plan of the Foundation Mining Path of Santa Barbara.

References

1. Lanzani, A.: I paesaggi italiani. vol. 21, Meltemi Editore srl (2003)
2. Casini, L.: Un patrimonio culturale senza frontiere? Aedon **2**, 1–3 (2018)
3. Pettenati, G.: I paesaggi culturali Unesco in Italia, FrancoAngeli (2019)
4. http://www.unesco.it/it/ItaliaNellUnesco/Detail/188
5. Palmi, P., Esposito, M., Prete, M.I.: Change in perspectives in cultural tourism: a sustainable managerial model for cultural thematic routes creating territorial value. In: Demartini, P., Marchegiani, L., Marchiori, M., Schiuma, G. (eds.) Cultural Initiatives for Sustainable Development. CMS, pp. 199–223. Springer, Cham (2021). https://doi.org/10.1007/978-3-030-65687-4_10
6. UNWTO. COVID-19: Putting people first. UNWTO (2020). https://www.unwto.org/tourism-covid-19
7. Balletto, G., Milesi, A., Ladu, M., Borruso, G.: A Dashboard for Supporting Slow Tourism in Green Infrastructures. A Methodological Proposal in Sardinia (Italy). Sustainability **12**(9), 3579 (2020)
8. https://www.beniculturali.it/mibac/export/MiBAC/sito-MiBAC/Contenuti/MibacUnif/Comunicati/visualizza_asset.html_611361093.html
9. Costa, S., Coles, R., Boultwood, A.: Landscape experience and the speed of walking. Landscapes in Flux, 124–128 (2015)
10. Notarstefano, G., Gristina, S.: Eco-sustainable routes and religious tourism: an opportunity for local development. The case study of Sicilian routes. In: Tourism in the Mediterranean Sea, Emerald Publishing Limited (2021)
11. Portale. www.camminiditalia.it
12. MiBACT. Piano Strategico di Sviluppo per il Turismo 2017–2022. Roma: MiBACT (2016). https://www.beniculturali.it/mibac/multimedia/MiBAC/documents/1481892223634_PST_2017_IT.pdf
13. Basile, G., Cavallo, A.: Rural identity, authenticity, and sustainability in Italian inner areas. Sustainability **12**(3), 1272 (2020)
14. Parco Geominerario Web Site. http://www.parcogeominerario.eu/ Accessed 15 Sept 2019
15. Balletto, G., Milesi, A., Ladu, M., Borruso, G.: Le reti per la reinvenzione del passato. Il caso del Cammino di Santa Barbara (Sardegna, Italia). In: Proceedings of the 23rd IPSAPA/ISPALEM International Scientific Conference Napoli (Italy), July 4–5, pp. 179–191 (2019)
16. Balletto, G., Milesi, A., Mundula, L., Borruso, G.: Wave, walk and bike tourism. The case of Sulcis (Sardinia-Italy). In: Gargiulo, C., Zoppi, C., (Eds.) Planning, Nature and Ecosystem Services, FedOAPress, Naples, Italy, pp. 881–892 (2019)
17. Ladu, M., Milesi, A., Borruso, G., Balletto, G.: Turismo lento nel Sulcis Iglesiente. Mappe di comunità per le sfide dello sviluppo turistico locale. In: Atti della XXIII Conferenza Nazionale ASITA, 12–14 November, Trieste, pp. 595–602 (2019)
18. Balletto, G., Milesi, A., Battino, S., Borruso, G., Mundula, L.: Slow tourism and smart community. The case of sulcis - iglesiente (Sardinia -Italy). In: Misra. S., et al. (eds) Computational Science and Its Applications – ICCSA 2019. ICCSA 2019. LNCS, vol. 11624. Springer, Cham (2019). https://doi.org/10.1007/978-3-030-24311-1_13

19. Ladu, M., Balletto, G., Milesi, A., Borruso, G.: Il ruolo delle tecnologie digitali nella promozione del turismo lento in Sardegna (Italia). Una proposta per il Cammino di Santa Barbara. In: Atti della XXIII Conferenza Nazionale SIU, Torino 17–18 Giugno 2021, Vol. 08 - Piani e politiche per una nuova accessibilità, Planum Publisher e Società Italiana degli Urbanisti, Roma Milano, pp. 80–89 (2021). https://doi.org/10.53143/PLM.C.821

20. Cammino Minerario di Santa Barbara Web Site. https://www.camminominerariodisantabarb ara.org/. Accessed 15 Sep 2019

21. Il sole24ore Web Site. https://www.ilsole24ore.com/art/santa-barbara-primo-posto-top-10-cammini-d-italia-ACdqaT9

22. Fondazione Cammino di Santa Barbara Il cammino minerario di Santa Barbara linee guida per lo sviluppo del prodotto turistico territoriale (2018). https://www.camminominerariodisan tabarbara.org/wp-content/uploads/2020/02/CMSB_Linee-Guida.pdf Accessed 11 May 2020

23. Balletto, G., Milesi, A., Fenu, N., Borruso, G., Mundula, L.: Military training areas as Semi-commons: The territorial Valorization of Quirra (Sardinia) from easements to ecosystem services. Sustainability **12**(2), 622 (2020)

Lightning Source UK Ltd.
Milton Keynes UK
UKHW020757020323
417918UK00006B/319